A. Barlotti (Ed.)

Finite Geometric Structures and their Applications

Lectures given at the
Centro Internazionale Matematico Estivo (C.I.M.E.),
held in Bressanone (Bolzano), Italy,
June 18-27, 1972

 Springer

FONDAZIONE
CIME
ROBERTO CONTI

C.I.M.E. Foundation
c/o Dipartimento di Matematica "U. Dini"
Viale Morgagni n. 67/a
50134 Firenze
Italy
cime@math.unifi.it

ISBN 978-3-642-10972-0   e-ISBN: 978-3-642-10973-7
DOI:10.1007/978-3-642-10973-7
Springer Heidelberg Dordrecht London New York

Printed on acid-free paper

Springer.com

CENTRO INTERNAZIONALE MATEMATICO ESTIVO

(C. I. M. E.)

2°Ciclo - Bressanone - dal 18 al 27 Giugno 1972

FINITE GEOMETRIC STRUCTURES AND THEIR APPLICATIONS

Coordinatore: Prof. A. BARLOTTI

CENTRO INTERNAZIONALE MATEMATICO ESTIVO

(C. I. M. E.)

R. C. BOSE

GRAPHS AND DESIGNS

Corso tenuto a Bressanone dal 18 al 27 Giugno 1972

These lectures were prepared for a C. I. M. E.
advanced Summer Institute held in 1972.
The undelying research was supported by the
National Science Foundation Grant GP 30958-X-

R. C. BOSE

CONTENTS

R. C. BOSE

CHAPTER I

BALANCED INCOMPLETE BLOCK DESIGNS

1. **Definition of Balanced Incomplete Block (BIB) designs.**
balanced incomplete block design may be defined to be an arrangement of v objects, called "treatments", into b subsets of these objects called "blocks", if the following conditions are satisfied:

(i)   Each block consists of k distinct treatments.

(ii)   Each treatment occurs in r different blocks.

(iii) Each pair of distinct treatments occur together in λ different blocks.

The positive integers v, b, r, k, λ are called the parameters of the design. These designs are of use in carrying out statistically controlled experiments. The parameter k is called the block size; and the paramenter r, the number of replications for the design.

Easy counting arguments show that the five parameters satisfy the following relations

(1.1.1) $$bk = vr, \quad \lambda(v - 1) = r(k - 1).$$

2. **Fisher's inequality for proper BIB designs.** Let θ and ϕ be two distinct treatments of a BIB design. Then θ occurs in r blocks. The λ blocks in which θ and ϕ occur together must form a subset of these r blocks. Hence

(1.2.1) $$\lambda \leqq r .$$

When the equality holds in (1.2.1), it follows from (1.1.1) that

(1.2.2) $$b = r = \lambda, \quad v = k.$$

In this case, each of the v treatments occurs in each of the b blocks.

This special case is known as the "randomized block design". A BIB design which is not a randomized block design, is called a proper BIB design. For a proper BIB design $r > \lambda$.

Given a BIB design, the incidence matrix N is defined as follows:

$$(1.2.3) \qquad\qquad N = (n_{ij}),$$

where $n_{ij} = 1$ if the i-th treatment occurs in the j-th block and $n_{ij} = 0$ if the i-th treatment does not occur in the j-th block. Thus, N is $v \times b$ matrix. Clearly

$$(1.2.4) \qquad \sum_{j=1}^{b} n_{ij}^2 = r, \quad \sum_{j=1}^{b} n_{ij} n_{\alpha j} = \lambda \text{ if } i \neq \alpha.$$

If N' is the transpose of N, then NN' is a square matrix of order v given by

$$(1.2.5) \qquad NN' = \begin{bmatrix} r & \lambda & \lambda & \cdots & \lambda \\ \lambda & r & \lambda & \cdots & \lambda \\ \lambda & \lambda & r & \cdots & \lambda \\ \cdots & \cdots & \cdots & \cdots & \cdots \\ \lambda & \lambda & \lambda & \cdots & r \end{bmatrix} = (r - \lambda)I_v + \lambda J_v$$

where I_v is the unit matrix of order v and J_v is a square matrix of order v each of whose elements is unity. It is readily seen that

$$(1.2.6) \qquad \det(NN') = rk(r - \lambda)^{v-1} .$$

It follows that for a proper BIB design $\det(NN') > 0$. Hence Rank $(NN') = v$. But the rank of N cannot exceed the number of columns in N. Hence

$$b \geqq \text{Rank } N$$
$$\geqq \text{Rank } NN'$$
$$= v.$$

Hence we have

Theorem (1.2.1). For a proper BIB design, the inequality

$$b \geqq v .$$

holds.

This inequality is due to Fisher (1940). The proof given here is due to Bose (1949). For example let $v = 16$, $b = 8$, $k = 6$, $\lambda = 1$. Then the conditions (1.1.1) are satisfied. Nevertheless, a BIB design with parameters 16, 8, 3, 6, 1 cannot exist since Fisher's inequality is violated.

3. Symmetric BIB designs. A BIB design is said to be symmetric if $v = b$ and in consequence $r = k$. We shall now prove

Theorem (1.3.1). For a symmetric BIB design with parameters $v = b$, $r = k$, λ any two blocks have exactly λ treatments in common.

If $r = \lambda$, the design is a randomized block design with $v = b = r = k = \lambda$ and the theorem is trivially true. Hence we can assume $r > \lambda$.

Now N is a $v \times v$ square matrix and since

$$\sum_{j=1}^{b} n_{ij} = r = k = \sum_{i=1}^{v} n_{ij} ,$$

we have

$$NJ_v = J_v N = rJ_v .$$

But from (1.2.5)

(1.3.1)
$$\begin{aligned} NN'N &= \{(r - \lambda)I_v + J_v\} N \\ &= N \{(r - \lambda)I_v + J_v\} \\ &= NNN' . \end{aligned}$$

Again from (1.2.6)

$$(\det N)^2 = \det(NN') > 0 .$$

Hence det $N \neq 0$ which shows that N is non-singular. Multiplying both side of (1.3.1) from the left by N^{-1} we have

$$N'N = NN' .$$

This is equivalent to the result to be established.

4. <u>The complementary of a BIB design</u>. Given a BIB design with parameters v, b, r, k, λ, we can form another design, called the complementary of the original, by taking in the j-th block of the complementary, those treatments which do not occur in the j-th block of the original. The number of blocks and the number of treatments in the complementary is the same as in the original design. Also the number of treatments in any block is $v - k$. Hence using the subscript "0" from the parameters of the complementary, we have

$$v_0 = v, \quad b_0 = b, \quad k_0 = v - k .$$

The treatment θ will occur in the j-th block of the complementary, if it does not occur in the j-th block of the original. Since there are r blocks of the original in which θ occurs, there are $b - r$ blocks of the original in which θ does not occur. This shows that

$$r_0 = b - r .$$

A pair of treatments θ and ϕ will occur together in the j-th block of the complementary, if neither of them occurs in the j-th block of the original. But θ and ϕ each occur in r block of the original of which λ are in common. Hence the number of blocks of the original in which either one or both of θ and ϕ occur is $2r - \lambda$. Thus, there are $b - 2r + \lambda$ blocks of the original in which neither θ nor ϕ occur.

$$\lambda_0 = b - 2r + \lambda .$$

We thus have the theorem

Theorem (1.4.1). If a BIB design has the parameters v, b, r, k, λ, the complementary BIB design has the paramenters

(1.4.1) $v_0 = v, \quad b_0 = b, \quad r_0 = b - r, \quad k_0 = v - k, \quad \lambda_0 = b - 2r + \lambda.$

A given BIB design and its complementary uniquely determine each other. For at least one of these, the block size $k \leq v/2$, where v is the number of treatments. Hence, when trying to construct BIB designs we need to obtain only those for which $k \leq v/2$.

5. The residual and the derived of a BIB design. Consider a symmetric BIB design D with parameters

$$v = b, \quad r = k, \quad \lambda,$$

where in virtue of (1.1.1),

$$\lambda(v - 1) = r(k - 1).$$

Choose any block of D as initial block and delete from D the initial block and the treatments contained in the initial block. We shall show that the remaining blocks form a BIB design, the residual design of D. Since one block and k treatments have been deleted, the number of treatments and blocks remaining is given by

$$v^* = v - k, \quad b^* = b - 1.$$

From Theorem (1.3.1), any two blocks have exactly λ treatments in common. Hence the number of deleted treatments in each of the b - 1 blocks retained is λ. Thus, the block size in the new design is

$$k^* = k - \lambda.$$

Finally, we note that treatments and pairs of treatments not occurring in the initial block remain undisturbed. Hence

$$r^* = r, \quad \lambda^* = \lambda.$$

The process by which the new design is obtained is called the process of block section, and the new design is called the residual of the original design. Hence we have the theorem

Theorem (1.5.1). From a symmetric BIB design D with parameters $v = b$, $r = k, \lambda$ we can obtain, by the process of block section, a new BIB design D^* (the residual of D), whose parameters are

(1.5.1) $v^* = v - k, \quad b^* = b - 1, \quad k^* = k - \lambda, \quad r^* = r, \quad \lambda^* = \lambda$.

Again starting from the symmetric BIB design with parameters $v = b$, $r = k, \lambda$ we can delete an initial block, and from the other blocks retain only the treatments contained in the initial block. Then the number of retained treatments and blocks is given by

$$v' = k, \quad b' = b - 1.$$

Also since any two blocks of the original design D have exactly λ treatments in common, we have in the new design

$$k' = \lambda.$$

Since each of the retained treatments is contained once in the initial deleted block, and also each pair of retained treatments occurs once in the initial deleted block, we have in the new design

$$r' = r - 1, \quad \lambda' = \lambda - 1.$$

The process by which the new design has been obtained is called the process of block intersection, and the new design is called the derived of the original design. Hence we have the theorem

Theorem (1.5.2). From the symmetric BIB design D with parameters
$v = b$, $r = k$, λ we can obtain by the process of block intersection a new
BIB design D' (the derived of D), whose parameters are

(1.5.2) $v' = k$, $b' = b - 1$, $k' = \lambda$, $r' = r - 1$, $\lambda' = \lambda - 1$.

6. Resolvable and affine resolvable BIB designs. We shall first
prove the following Lemma:

Lemma (1.6.1). If a BIB design (not a randomized block design has
parameters v, b, r, k, λ and the number of blocks b is divisible by r,
then Fisher's inequality can be refined to

(1.6.1) $b \geq v + r - 1$.

From (1.1.1)

$$\frac{b - r}{v - 1} = \frac{r - \lambda}{k}$$

and

$$rk - v\lambda = r - \lambda.$$

Let $b = nr$, $v = nk$, then

$$\frac{r - \lambda}{k} = r - \frac{v\lambda}{k} = r - n\lambda.$$

Since $r > \lambda$, and $r - n\lambda$ is integral we have $r - n\lambda \geq 1$. Hence

$$\frac{b - r}{v - 1} \geq 1$$

or

$$b \geq v + r - 1.$$

A BIB design with parameters v, b, r, k, λ is said to be resolvable
if it is possible to separate the b blocks into r sets

(S_0), (S_1),..., (S_{r-1}) with each set consisting of b/r blocks, so that every treatment occurs exactly once in the blocks of any set (S_i), i = 0, 1, 2,..., r - 1. The blocks of each set (S_i) are said to constitute a complete replication. If further any two blocks of different sets or replications have the same number m of treatments in common, the design is said to be **affine resolvable**.

Since b is divisible by r for a resolvable design, the inequality (1.6.1) holds. It would be of interest to study the conditions under which the equality holds in (1.6.1).

Let the blocks belonging to the set (S_i) be

$$B_{i1}, B_{i2}, \ldots, B_{in} \quad (i = 0, 1, \ldots, r-1)$$

where b = nr and v = nk. Let us take any particular block, say the block B_{01} of the set (S_0). Let ℓ_{ij} be the number of treatments common to the block B_{01} and the block B_{ij} of the set (S_i), for i = 1, 2, ..., r-1, j = 1, 2, ..., n. Let m denote the mean and s^2 the variance of n(r-1) quantities ℓ_{ij}.

Each of the k treatments occurring in B_{01} occurs in the design r times. If a treatment occurs in B_{01}, it cannot occur in the other blocks of the set (S_0). Hence it occurs just r - 1 times among the blocks of the sets (S_1), (S_2), ..., (S_{r-1}). Hence

$$\Sigma \, \ell_{ij} = k(r-1).$$

$$\therefore m = \frac{k}{n} = \frac{k^2}{v}.$$

Again the k(k - 1)/2 pairs involved in B_{01} each appear $\lambda - 1$ times in the sets (S_1), (S_2), ..., (S_{r-1}). Hence

$$\tfrac{1}{2} \Sigma \ell_{ij} (\ell_{ij} - 1) = \tfrac{1}{2} (\lambda - 1) k (k - 1).$$

$$\therefore \Sigma \ell_{ij}^2 = k \{(r - 1) + (\lambda - 1)(k - 1)\}.$$

Now from (1.1.1)

$$\lambda = \frac{r(k-1)}{nk-1}.$$

Hence

$$\Sigma \ell_{ij}^2 = \frac{k\{(nk-1)(r-k) + r(k-1)^2\}}{nk-1}$$

$$\therefore s^2 = \frac{\Sigma(\ell_{ij} - m)^2}{n(r-1)}$$

$$= \frac{k^2(n-1)\{r(n-1) - (nk-1)\}}{n^2(r-1)(nk-1)}.$$

Hence remembering that $b = nr$, $v = nk$, we have

$$s^2 = \frac{k(v-k)(b-v-r+1)}{n^2(r-1)(v-1)}.$$

Since $s^2 \geq 0$, we have an alternative proof of the inequality

$$b \geq v + r - 1.$$

If $b = v + r - 1$, then $s = 0$. Hence $\ell_{ij} = m$ for all values of i and j. This shows that the block B_{01} has exactly m treatments in common with each block of the sets (S_1), ..., (S_{r-1}). Since the choice of B_{01} was arbitrary, this shows that any two blocks of different sets have exactly m treatments in common; i.e., the design is affine resolvable. Also since m must be integral, k^2/v is integral.

Conversely if the original BIB design is affine resolvable then ℓ_{ij} must be constant for $i = 1, 2, \ldots, r-1$, $j = 1, 2, \ldots, n$. Hence $s^2 = 0$ or $b = v + r - 1$. The constant value of ℓ_{ij} is $m = k^2/v$. We can sum up our results in the following theorem:

R. C. BOSE

Theorem (1.6.1). If a BIB design with parameters v, b, r, k, λ is resolvable

(1.6.2) $$b \geq v + r - 1.$$

If further the design is affine resolvable, then the equality holds in (1.6.2). Conversely if for a resolvable design, the equality holds in (1.6.2), then the design must be affine resolvable. Any two blocks of an affine resolvable design which belong to different replications have exactly k^2/v treatments in common. Hence, this number must be integral.

7. Use of finite geometries for the construction of BIB designs.

(a) Consider a finite projective plane of order $s = p^n$. We may identify the points of the plane with treatments, and the lines of the plane with blocks. Since each line contains $s + 1$ points, through each point pass $s + 1$ lines ($s = p^n$), and each pair of points is joined by exactly one line, we get a symmetric BIB design with paramenters

(1.7.1) $$v = b = s^2 + s + 1, \quad r = k = s + 1, \quad \lambda = 1.$$

From any finite affine plane of order $s = p^n$, we can similarly obtain a BIB design with parameters

(1.7.2) $$v = s^2, \quad b = s^2 + s, \quad r = s + 1, \quad k = s, \quad \lambda = 1,$$

by identifying the points of the plane with treatments of the design, and the lines of the plane with the blocks of the design.

(b) In the same way we can obtain BIB designs by using the N-dimensional finite projective space $PG(N, p^n)$ and the N-dimensional affine projective space $EG(N, p^n)$. The function

(1.7.3) $$(N, m, s) = \frac{(s^{N+1} - 1)(s^N - 1) \cdots (s^{N-m+1} - 1)}{(s^{m+1} - 1)(s^m - 1) \cdots (s - 1)}$$

denotes the number of m-flats in $PG(N,p^n)$, where $s = p^n$.

By taking the points of $PG(N,p^n)$ as treatments, and the m-flats of it as blocks, we get the BIB design with parameters v, b, r, k, λ where $p^n = s$, and

$$(1.7.4) \begin{cases} v = \phi(N,0,s) = \dfrac{s^{N+1}-1}{s-1} \, , \\[2ex] b = \phi(N,m,s) = \dfrac{(s^{N+1}-1)(s^N-1) \cdots (s^{N-m+1}-1)}{(s^{m+1}-1)(s^m-1) \cdots (s-1)} \, , \\[2ex] r = \phi(N-1,m-1,s) = \dfrac{(s^N-1)(s^{N-1}-1) \cdots (s^{N-m+1}-1)}{(s^m-1)(s^{m-1}-1) \cdots (s-1)} \, , \\[2ex] k = \phi(m,0,s) = \dfrac{s^{m+1}-1}{s-1} \, , \\[2ex] \lambda = \phi(N-2,m-2,s) = \dfrac{(s^{N-1}-1)(s^{N-2}-1) \cdots (s^{N-m+1}-1)}{(s^{m-1}-1)(s^{m-2}-1) \cdots (s-1)} \, . \end{cases}$$

The design (1.7.1) is then the special case N = 2, m = 1.

Again taking the points of $EG(N,p^n)$ as treatments, and m-flats as blocks, we get the BIB design with parameters v, b, r, k, λ where $s = p^n$ and

$$(1.7.5) \begin{cases} v = s^N \, , \\[2ex] b = s^{N-m} \phi(N-1,m-1,s) = \dfrac{s^{N-m}(s^N-1) \cdots (s^{N-m+1}-1)}{(s^m-1)(s^{m-1}-1) \cdots (s-1)} \, , \\[2ex] r = \phi(N-1,m-1,s) = \dfrac{(s^N-1)(s^{N-1}-1)(s^{N-m+1}-1)}{(s^m-1)(s^{m-1}-1) \cdots (s-1)} \, , \\[2ex] k = s^m \, , \\[2ex] \lambda = \phi(N-2,m-2,s) = \dfrac{(s^{N-1}-1)(s^{N-2}-1) \cdots (s^{N-m+1}-1)}{(s^{m-1}-1)(s^{m-2}-1) \cdots (s-1)} \, . \end{cases}$$

The design (1.7.2) is the special case N = 2, m = 1.

(c) BIB designs can also be obtained by using non-linear curves and surfaces. We shall illustrate this by a single example.

Let Q_2 be a non-degenerate conic in $PG(2,2^n)$. Then there are $s+1$ points on Q_2 where $s = 2^n$. It is known [Bose (1947 a), Qvist (1952)] that the $s+1$ tangents to Q_2 all pass through a point O, the nucleus of polarity of Q_2. The remaining s^2 lines can be divided in two sets viz. the $s(s+1)/2$ intersectors each meeting Q_2 in two points, and the $s(s-1)/2$ non-intersectors, which do not meet Q_2. Let the s^2-1 points not on Q_2 and other than the nucleus of polarity be called retained points. Let us study the configuration of the retained points and the non-intersectors. We shall identify our treatments with the non-intersectors and our blocks with the retained points, a treatment being contained in a block, if the corresponding non-intersector passes through the corresponding retained point. Clearly

$$v = s(s-1)/2, \quad b = s^2-1, \quad k = s/2.$$

If a treatment is contained in r blocks, then the corresponding non-intersector must pass through r retained points. But each non-intersector passes through $s+1$ points, all of which are retained points. Hence $r = s+1$.

Finally any two non-intersectors have exactly one point in common (which is a retained point). Hence $\lambda = 1$. We thus get a BIB design with parameters

(1.7.6) $\quad v = s(s-1)/2, \quad b = s^2-1, \quad r = s+1, \quad k = s/2, \quad \lambda = 1$

where $s = 2^n$.

8. **Non-existence theorems for symmetric BIB designs.** The conditions

$$bk = vr, \quad \lambda(v-1) = r(k-1),$$

R. C. BOSE

and Fisher's inequality

$$b \geqq v \, ,$$

are necessary but not sufficient conditions for the existence of a BIB design with parameters v, b, r, k, λ. We can derive further necessary conditions in certain special cases, but no sufficient conditions are known for the general case. We shall now prove:

Theorem (1.8.1). For a (proper) symmetric BIB design in which the number of treatments v is even, $r - \lambda$ must be a perfect square.

Let N be the incidence matrix of the design, From (1.2.6)

$$\det (NN') = r^2 (r - \lambda)^{v-1}.$$

However for symmetric designs

$$\det (NN') = (\det N)^2.$$

Since v is even $r^2 (r - \lambda)^{v-1}$ can be a perfect square only if $r - \lambda$ is a perfect square. This proves the theorem [Schützenberger (1949)]. For example, the following symmetric BIB designs are impossible:

$$v = b = 22, \quad r = k = 7, \quad \lambda = 2,$$
$$v = b = 46, \quad r = k = 10, \quad \lambda = 2.$$

The case when v is odd requires the use of more sophisticated methods. The first result in this direction was obtained by Bruck and Ryser (1949), who showed

Theorem (1.8.1). If $s \equiv 1 \pmod 4$ or $2 \pmod 4$, a necessary condition for the existence of a projective plane of order s(or equivalently the BIB design (1.7.1) is that if p is any odd prime dividing the square free part of s, then $p \not\equiv 3 \pmod 4$.

Example. A projective plane of order 6 or 14 cannot exist.

R. C. BOSE

Bruck and Ryser's result was generalized by Shrikhande (1950) and by
Chowla and Ryser (1950). The latter proved

Theorem (1.8.2). A necessary condition for the existence of a symmetric BIB design with parameter $v = b$, $r = k$, λ; when v is odd is that
the diophantine equation

$$x^2 = (r - \lambda)y^2 + (-1)^{\frac{v-1}{2}} \lambda z^2$$

has a solution in intergers, not all zero.

From this we can deduce the following result due to Shrikhande and
Raghawarao (1964).

Theorem (1.8.3). Let u be the square free part of $r - \lambda$ and t the
square free part of λ. Then for the existence of a symmetric BIB design
with parameters $v = b$, $r = k$, it is necessary that.

(a) for all odd primes p dividing u but not t, $((-1)^{\frac{v-1}{2}} t | p) = 1$

(b) for all odd primes p dividing t but not u, $(u | p) = 1$

(c) for all odd primes p dividing both u and t,

$(- (-1)^{\frac{v-1}{2}} u_0 t_0 | p) = 1$, where $u = u_0 p$ and $t = t_0 p$.

Here $(m | p)$ is the Legendre function defined by $(m | p) = 1$, -1 or 0
according as the residue class to which m belongs is a quadratic residue,
non-quadratic residue or the null class. Bruck and Ryser's theorem follows as a special case from (a).

For example BIB designs with the following parameters are non-exist-
ent:

(i) $v = b = 29$, $r = k = 8$, $\lambda = 2$,

(ii) $v = b = 141$, $r = k = 21$, $\lambda = 3$,

(iii) $v = b = 43$, $r = k = 15$, $\lambda = 5$.

The non-existence of the above designs follows by using parts (a),
(b), (c) respectively of the Theorem (1.8.3).

9. **Final remarks**. It is not possible to pursue in the limited space
available the various methods of construction and other properties of BIB
designs. The interested reader may refer to the following papers and
books: Bose [(1939), (1942 a), (1942 b), (1947 b), (1963 a)], Bose and
Shrikhande (1960 a), Fisher [(1940, (1942)], Hanani [(1961), (1965)], Hall
(1967), Mann (1949), Rao [(1946), (1961)], Ray-Chaudhuri and Wilson (1971),
Wilson [(1972 a), (1972 b)].

CHAPTER II

GROUP DIVISIBLE DESIGNS

1. **Definition of group divisible (GD) designs**. **Relations between**
parameters. Suppose there are $v = mn$ objects or treatments, which are
divided into m groups each with n treatments. A group divisible (GD)
design is then an arrangement of the v treatments into b sets or blocks
for which the following conditions are satisfied:

(i)   Each block consists of k distinct treatments.

(ii) Each treatment occurs in r different blocks.

(iii) Each pair of distinct treatments which belong to the same
group occur together in λ_1 blocks, whereas a pair of distinct treatments
which do not belong to the same group occur together in λ_2 blocks.

The scheme showing the division of the treatments into groups is
called the association scheme of the design. The association scheme will
usually be written as an $n \times m$ rectangular arrangement of the treatments,

in which treatments belong to a group appear in the same column of the scheme.

The integers v, b, r, k, m, n, λ_1, λ_2 are said to be the parameters of the design.

The parameters v, b, r, k, λ_1, λ_2, m, n are connected by three relations, so that only five are free. Clearly

(2.1.1) $$v = mn, \quad bk = vr .$$

Any given treatment occurs in r blocks. Since each of these blocks contains k - 1 other treatments, there are r(k - 1) pairs of which one member is θ. But θ must form λ_1 pairs with each of the n - 1 treatments belonging to the same group as θ, and λ_2 pairs with each of the n(m - 1) treatments not in the same group as θ. Hence

(2.1.2) $$(n - 1)\lambda_1 + n(m - 1)\lambda_2 = r(k - 1) .$$

It is also easy to see that

$$r \geq \lambda_1, \quad r \geq \lambda_2 .$$

2. **Classification of group divisible design.** Let n_{ij} = 1 or 0 according as the i-th treatment does or does not occur in the j-th block. Then the matrix

$$N = (n_{ij})$$

is defined to be the incidence matrix of the design. From the conditions satisfied by the design, it is readily seen that

(2.2.1) $$\sum_{j=1}^{v} n_{ij}^2 = r, \quad \sum_{j=1}^{v} n_{ij} n_{uj} = \lambda_1 \text{ or } \lambda_2$$

according as the i-th and u-th treatments do or do not belong to the same group.

In numbering the treatments we shall follow the convention that the th group consists of the treatments number $n(\ell-1)+1$, $n(\ell-1)+2$, ..., $n\ell$. It follows from (2.2.1) that

$$NN' = \begin{bmatrix} A & B & \cdots & B \\ B & A & \cdots & B \\ \cdots & \cdots & \cdots & \cdots \\ B & B & \cdots & A \end{bmatrix}$$

where N' is the transpose of N, and A and B are $n \times n$ matrices defined by

$$A = \begin{bmatrix} r & \lambda_1 & \cdots & \lambda_1 \\ \lambda_1 & r & \cdots & \lambda_1 \\ \cdots & \cdots & \cdots & \cdots \\ \lambda_1 & \lambda_1 & \cdots & r \end{bmatrix} \qquad B = \begin{bmatrix} \lambda_2 & \lambda_2 & \cdots & \lambda_2 \\ \lambda_2 & \lambda_2 & \cdots & \lambda_2 \\ \cdots & \cdots & \cdots & \cdots \\ \lambda_2 & \lambda_2 & \cdots & \lambda_2 \end{bmatrix}$$

To find the characteristics roots of NN' we have to evaluate $|NN' - I\theta|$ where I is the unit natrix of order mn. After some reduction we obtain

(2.2.2) $|NN' - I\theta| = (rk - \theta)(rk - v\lambda_2 - \theta)^{m-1} (r - \lambda_1 - \theta)^{m(n-1)}$.

Hence we have the following theorem:

Theorem (2.2.1). If N is the incidence matrix of a GD design with parameters v, b, r, k, λ_1, λ_2, m, n, the characteristic roots of NN' are rk, $rk - v\lambda_2$ and $r - \lambda_1$ with multiplicities 1, $m-1$ and $m(n-1)$ respectively.

Corollary. For a GD design $rk - v\lambda_2 \geq 0$.

This follows because the characteristic roots of NN' must be non-negative.

We can divide GD designs into three exhaustive and mutually exclusive classes

   (a) Singular GD designs characterized by $r = \lambda_1$,

   (b) Semi-regular GD designs characterized by $r > \lambda_1$, $rk - v\lambda_2 = 0$,

(c) Regular GD designs characterized by $r > \lambda_1$, $rk - v\lambda_2 > 0$.

3. <u>Singular GD designs</u>. Consider a BIB design with parameters v^*, b^*, r^*, k^*, λ^*. If we replace each treatment by a group of n treatments we get $v = nv^*$ treatments divided into v^* groups, where each group corresponds to one of the original treatments. Two treatments belonging to the same group now occur together r^* times and two treatments belonging to different groups occur together λ^* times. We thus get a GD design with parameters

$$v = nv^*, \quad b = b^*, \quad r = r^*, \quad k = nk^*$$
$$\lambda_1 = r^*, \quad \lambda_2 = \lambda^*, \quad m = v^*, \quad n = n$$

which is a singular GD design since $r - \lambda_1 = 0$.

Conversely consider a singular GD design with parameters v, b, r, k, λ_1, λ_2, m, n, where $r = \lambda_1$. Let θ and ϕ be any two treatments belonging to the same group. Now θ occurs in r blocks and since $r = \lambda_1$, ϕ must occur in each of these r blocks and nowhere else. Hence if a treatment occurs in a certain block, every treatment belonging to the group occurs in that block. Let each group of treatments be replaced by a single treatment in the design, then there are $v^* = m$ treatments in the new design, and because any two treatments belonging to different groups occur together λ_2 times in the original GD design, the new design is a BIB design with parameters

$$v^* = m, \quad b^* = b, \quad r^* = r, \quad k^* = k/n, \quad \lambda^* = \lambda_2.$$

We thus get

<u>Theorem (2.3.1)</u>. If in a BIB design with parameters v^*, b^*, r^*, k^*, λ^* each treatment is replaced by a group of n treatments we get a GD design with parameters

R. C. BOSE

(2.3.1)

$$v = nv^*, \ b = b^*, \ r = r^*, \ k = nk^*, \ \lambda_1 = r^*, \ \lambda_2 = \lambda^*, \ m = v^*, \ n = n.$$

Conversely every singular GD design is obtainable in this way from a corresponding BIB design.

Corollary. For a singular GD design $b \geqq m$.

This follows from the inequality $b^* \geqq v^*$ which holds for the BIB design from which the GD design has been obtained.

4. Semi-regular GD designs. For a semi-regular GD design we have by definition

(2.4.1) $$r - \lambda_1 > 0, \quad rk - v_2 = 0.$$

Hence from (2.1.1) and (2.1.2) we have

(2.4.2) $$r + (n-1)\lambda_1 = n\lambda_2.$$

We shall now prove

Theorem (2.4.1). For a semi-regular GD design k is divisible by m. If $k = cm$, then every block must contain c treatments for every group.

Let e_j treatments from the first group occur in the jth block $(j = 1, 2, \ldots, b)$. Then

(2.4.3) $$\sum_{j=1}^{b} e_j = nr, \quad \sum_{j=1}^{b} e_j(e_j - 1) = n(n-1)\lambda_1$$

since each treatment from the first group occurs in r blocks, and every pair of treatments from the first group occurs in λ_1 blocks. From (2.4.2) and (2.4.3)

$$\sum_{j=1}^{b} e_j^2 = n^2 \lambda_2.$$

R. C. BOSE

From (2.1.1)

$$\bar{e} = \frac{1}{b} \sum_{j=1}^{b} e_j = \frac{nr}{b} = \frac{k}{m} .$$

Hence

$$\sum_{j=1}^{b} (e_j - \bar{e})^2 = n^2 \lambda_2 - \frac{bk^2}{m^2} = 0$$

from (2.1.1) and (2.4.1). Therefore,

$$e_1 = e_2 = \ldots = e_b = \bar{e} = \frac{k}{m} .$$

Since e_j must be integral, k must be divisible by m. If $k = cm$ then $e_j = c$ ($j = 1, 2, \ldots, b$). The same argument applies to treatments of any other group. This proves our theorem.

It follows from Theorem (2.2.1), that for a semi-regular GD design

$$\text{Rank } NN' = v - m + 1.$$

Now

$$b \geq \text{Rank } N$$

$$= \text{Rank } NN'$$

$$= v - m + 1.$$

If the design is resolvable then the sum of the column vectors in N, which correspond to a complete replication is the $v \times 1$ column vector all of whose elements are unity. In this case the rank of N cannot exceed $b - r + 1$. So for a resolvable semi-regular GD design

$$b - r + 1 \geq v - m + 1.$$

We then have

Theorem (2.4.2). For a semi-regular GD design with parameters $v, b, r, k, \lambda_1, \lambda_2, m, n$

$$b \geq v - m + 1.$$

If the design is resolvable, the inequality can be sharpened to

$$b \geq v - m + r.$$

5. <u>Regular GD designs</u>. For a regular GD design

$$r > \lambda_1, \quad rk - v\lambda_2 > 0.$$

From Theorem (2.2.1), NN' is non-singular. Hence

$$v = \text{Rank } NN' = \text{Rank } N \leq b.$$

If the design is resolvable

$$\text{Rank } N \leq b - r + 1.$$

Hence we have

<u>Theorem (2.5.1)</u>. For a regular GD design with parameters
$v, b, r, k, \lambda_1, \lambda_2, m, n$

$$b \geq v.$$

If the design is resolvable this inequality may be sharpened to

$$b \geq v + r - 1.$$

6. <u>Necessary conditions for the existence of symmetrical regular GD</u>
<u>designs</u>. A GD design is said to be symmetrical if $b = v$ and in conse-
quence $r = k$. Consider a symmetrical regular GD design with parameters
$v, b, r, k, \lambda_1, \lambda_2, m, n$, where

(2.6.1) $$v = b = mn, \quad r = k$$

(2.6.2) $$(n-1)\lambda_1 + n(m-1)\lambda_2 = r(r-1)$$

(2.6.3) $$Q = r - \lambda_1 > 0, \quad P = r^2 - v\lambda_2 > 0.$$

From (2.2.2)

$$|N|^2 = |NN'| = r^2 P^{m-1} Q^{m(n-1)}.$$

R. C. BOSE

It follows that $P^{m-1} Q^{m(n-1)}$ is a perfect square. Hence we have the theorem

Theorem (2.6.1). A necessary condition for the existence of a symmetrical regular GD design with parameters v, b, r, k, λ_1, λ_2, m, n is that $P^{m-1} Q^{m(n-1)}$ is a perfect square, where P and Q are given by (2.6.3).

Corollary. If m is even P must be a perfect square, and if m is odd and n is even then Q must be a perfect square.

We give below a table of some symmetrical GD designs whose impossibility can be proved by using Theorem (2.6.1).

Table (2.6.1)

Some impossible symmetrical regular GD designs

Ref. No.	v=b	r=k	m	n	λ_1	λ_2	P	Q
(1)	44	7	22	2	0	1	5	7
(2)	92	10	46	2	0	1	8	10
(3)	56	8	28	2	2	1	8	6
(4)	88	10	22	4	2	1	12	6
(5)	20	5	10	2	2	1	5	3
(6)	18	7	2	9	3	2	13	4

7. Final remarks. Further theorems on the impossibility of GD designs, analogous to the Chowla-Ryser (1950) and Shrikhande-Raghavarao (1964) theorems for BIB designs can be obtained by using Hilbert norm residue symbols [Bose and Connor (1952)].

For methods of constructing and additional properites of GD designs the interested reader is referred to Bose, Shrikhande and Bhattacharya (1953).

R. C. BOSE

CHAPTER III

PARTIALLY BALANCED DESIGNS AND ASSOCIATION SCHEMES

1. <u>Definition of partially balanced association schemes and partially balanced incomplete block (PBIB) designs</u>. Given v objects or
treatments 1, 2, ..., v a relation satisfying the following conditions is
said to be an association scheme with m-classes:

(a) Any two treatments are either 1st, 2nd, ..., or m-th associates,
the relation of association being symmetrical, i.e., if the treatment α
is the i-th associate of the treatment β, then β is the i-th associate of
the treatment α.

(b) Each treatment has n_i, i-th associates, the number n_i being
independent of α.

(c) If any two treatments are i-th associates then the number of
treatments which are j-th associates of α and k-th associates of β is p^i_{jk}
and is independent of the pair of i-th associates α and β.

The numbers

(3.1.1) $v, n_i, p^i_{jk} \quad (i, j, k = 1, 2, .., m),$

are the parameters of the association scheme.

If we have an association scheme with m classes, then we get a PBIB
design with r replications and b blocks based on the association scheme,
if we can arrange the v treatments in b blocks such that

(i)   Each block contains k treatments (all different).

(ii)   Each treatment is contained in r blocks.

(iii) If two treatments α and β are i-th associates, then they
occur together in λ_i blocks, the number λ_i being independent of the particular pair of i-th associates α and $\beta(i = 1, 2, ..., m)$.

For a PBIB design based on any association scheme, the parameters of the scheme may be called parameters of the first kind, and the additional parameters

(3.1.2) $\qquad$ b, r, k, λ_i (i = 1, 2, ..., m),

may be called parameters of the second kind. Clearly

(3.1.3) $\qquad$ $vr = bk$, $n_1 + n_2 + \ldots + n_m = v - 1$,

(3.1.4) $\qquad$ $n_1\lambda_1 + n_2\lambda_2 + \ldots + n_m\lambda_m = r(k-1)$.

2. **Relations between the parameters of association schemes.** By definition the number p_{jk}^i is independent of which pair α, β of i-th associates we start with. Consider the pair β, α; we see at once that

(3.2.1) $\qquad$ $p_{jk}^i = p_{kj}^i.$

The following further relations are easy to prove:

(3.2.2) $\qquad$ $\sum_{i=1}^{m} n_i = v - 1;$

(3.2.3) $\qquad$ $\sum_{k=1}^{m} p_{jk}^i = n_j \qquad$ if $i \neq j$

$\qquad\qquad\qquad\qquad = n_j - 1$ if $i = j;$

(3.2.4) $\qquad$ $n_i p_{jk}^i = n_j p_{ik}^i = n_k p_{ij}^k.$

These relations were proved by Bose and Nair (1939), in their paper introducing the PBIB designs. These are all the realtions in case m = 2 but for m ⩾ 3 further relations were discovered by Bose and Mesner (1959).

It is useful to make a convention that each treatment is its own zero-th associate and of no other treatments. Then clearly we must take,

(3.2.5) $\qquad$ $n_0 = 1;$

$$(3.2.6) \qquad p_{ji}^0 = p_{ij}^0 = 0 \quad \text{if } i \neq j,$$

$$= n_j \quad \text{if } i = j;$$

$$(3.2.7) \qquad p_{k0}^i = p_{0k}^i = 0 \quad \text{if } i \neq k,$$

$$= 1 \quad \text{if } i = k.$$

We can now write

$$(3.2.8) \qquad \sum_{i=0}^{m} n_i = v, \quad \sum_{k=0}^{m} p_{jk}^i = n_j$$

for i, j, k = 0, 1, ... m. It should also be noted that (3.2.4) remains valid if one or more of i, k, j is zero.

Also for a PBIB design based on the association scheme we must have

$$(3.2.9) \qquad \sum_{i=0}^{m} n_i \lambda_i = rk, \quad \text{where } \lambda_0 = r,$$

For a two class association scheme the values of the parameters p_{jk}^i (i, j, k = 1, 2) may conveniently be written in the form of two symmetric matrices

$$(3.2.10) \qquad (p_{jk}^1) = \begin{pmatrix} p_{11}^1 & p_{12}^1 \\ p_{21}^1 & p_{22}^2 \end{pmatrix}, \quad (p_{jk}^2) = \begin{pmatrix} p_{11}^2 & p_{12}^2 \\ p_{21}^2 & p_{22}^2 \end{pmatrix}.$$

The definition given in paragraph 1, for association schemes is not minimal, i.e., the constancy of some of the parameters can be deduced from others. In particular for two class association schemes, Bose and Clatworthy (1955) proved that the constancy of v, n_1, p_{11}^1, p_{11}^2 guarantees the constancy of all the other parameters of a two class association scheme.

3. <u>Some examples of two class association schemes</u>. We shall give below some examples of two class association schemes. This enumeration

R. C. BOSE

is for illustrative purposes and is not exhaustive. For examples of PBIB
designs based on them, see Bose (1963 b).

(a) <u>The group divisible (GD) association scheme</u>. In this case there
are mn treatments, which are divided into m groups of n treatments each.
Two treatments belonging to the same group are first associates, and two
treatments belonging to different groups are second associates. The
association scheme can be exhibited by writing down the mn treatments in
the form of a rectangular array, the treatments of the same group occupy-
ing the same column. It is readily seen that the parameters of the association scheme so obtained are

(3.3.1) $v = mn, \ n_1 = n - 1, \ n_2 = n(m - 1),$

(3.3.2) $(p_{jk}^1) = \begin{pmatrix} n - 2 & 0 \\ 0 & n(m - 1) \end{pmatrix}, \ (p_{jk}^2) = \begin{pmatrix} 0 & n - 1 \\ n - 1 & n(m - 2) \end{pmatrix}.$

For example, let m = 4, n = 3. The corresponding GD association
scheme is

$$\begin{array}{cccc} 1 & 2 & 3 & 4 \\ 5 & 6 & 7 & 8 \\ 9 & 10 & 11 & 12. \end{array}$$

The first associates of the treatment 1 are 5 and 9, and the second
associates are 2, 3, 4, 6, 7, 8, 10, 11, 12.

GD designs have already been considered in Chapter II. They are now
seen to be a special class of PBIB designs, viz., those which are based on
the GD association scheme.

(b) <u>The triangular association scheme</u>. We take an m × m square, and
fill-in the m(m - 1)/2 positions above the leading diagonal by different
treatments, taken in any order. The positions in the leading diagonal
are left blank, while positions below this diagonal are filled so that

the scheme is symmetrical with respect to the diagonal. Two treatments

in the same row (or same column) are first associates. Two treatments

which do not occur in the same row or same column are second associates.

It is readily verified that the parameters of the association scheme so

obtained are

(3.3.3) $v = m(m-1)/2, \quad n_1 = 2m-4, \quad n_2 = (m-2)(m-3)/2,$

(3.3.4)
$$(p_{jk}^1) = \begin{pmatrix} m-2 & m-3 \\ m-3 & (m-3)(m-4)/2 \end{pmatrix}, \quad (p_{jk}^2) = \begin{pmatrix} 4 & 2m-8 \\ 2m-8 & (m-4)(m-5)/2 \end{pmatrix}.$$

This scheme is called the triangular association scheme.

As an illustration take m = 5. The association scheme is then

*	1	2	3	4
1	*	5	6	7
2	5	*	8	9
3	6	8	*	10
4	7	9	10	*

Two treatments which are in the same row or same column are first

associates. Two treatments which do not occur in the same row or column

are second associates.

(c) The singly linked block (SLB) association scheme. Consider a

balanced incomplete block (BIB) design D with b treatments, v blocks, k

replications, block size r and $\lambda = 1$, i,e, every pair of treatments occurs

in exactly one block. Then

$$bk = vr, \quad b-1 = k(r-1).$$

Consider v new treatments each corresponding to one block of D. Two

of these new treatments will be called first associates if the correspond-

ing blocks of D have a common treatment and second associates if the corresponding blocks of D have no common treatment. Shrikhande (1952) has shown that this association relation satisfies the conditions (a), (b), (c) of paragraph 1 with parameters,

$$(3.3.5) \quad v = k(kr - k + 1)/r, \quad n_1 = r(k - 1), \quad n_2 = (k - r)(r - 1)(k - 1)/r,$$

$$(3.3.6) \quad (p_{jk}^1) = \begin{pmatrix} (k-2) + (r-1)^2 & (r-1)(k-r) \\ (r-1)(k-r) & (r-1)(k-r)(k-r-1)/r \end{pmatrix},$$

$$(3.3.7) \quad (p_{jk}^2) = \begin{pmatrix} r^2 & r(k-r-1) \\ r(k-r-1) & \frac{(k-r)(r-1)(k-1)}{r} - r(k-r-1) - 1 \end{pmatrix}.$$

This association scheme is defined to be an SLB scheme. Every BIB design with $\lambda = 1$ gives rise to such a scheme.

(d) The Latin square (L_r) association scheme. Consider $v = k^2$ treatments which may be set forth in a $k \times k$ scheme. Thus if $k = 4$ and the treatments are $1, 2, \ldots, 16$, we have the scheme

(3.3.8)

1	2	3	4
5	6	7	8
9	10	11	12
13	14	15	16

For the case $r = 2$, we define two treatments as first associates if they occur in the same row or column of the square scheme, and second associates otherwise. The association scheme so defined may be called the L_2 association scheme. The parameters of the L_2 scheme are

$$(3.3.9) \quad v = k^2, \quad n_1 = 2(k-1), \quad n_2 = (k-1)^2,$$

$$(3.3.10) \quad (p_{jk}^1) = \begin{pmatrix} k-2 & k-1 \\ k-1 & (k-1)(k-2) \end{pmatrix}; \quad (p_{jk}^2) = \begin{pmatrix} 2 & 2(k-2) \\ 2(k-2) & (k-2)^2 \end{pmatrix}.$$

In the general case $2 \leq r \leq k + 1$, we take a set of $r - 2$ mutually orth-

ogonal Latin squares (if such a set exists). For an L_r association
scheme we then define two treatments to be first associates if they occur
together in the same row or column of the square scheme, or if they correspond to the same symbol of one of the Latin squares. Otherwise we define them to be second associates. For example if $k = 4$, $r = 4$ and we
take the Latin squares

$$[L_1] \qquad\qquad [L_2]$$

(3.3.11)

1	2	3	4
2	1	4	3
3	4	1	2
4	3	2	1

1	2	3	4
3	4	1	2
4	3	2	1
2	1	4	3

then the first associates of the treatment 7 and 5, 6, 8, 3, 11, 15, 4,
10, 13, 1, 12, 14 because the treatment 7 corresponds to the symbol 4 in
$[L_1]$ and the symbol 1 in $[L_2]$. The parameters of the L_r association
scheme are given by

$$(3.3.12) \qquad v = k^2, \quad n_1 = r(k-1), \quad n_2 = (k-1)(k-r+1),$$

$$(3.3.13) \qquad (p^1_{jk}) = \begin{pmatrix} (k-2)+(r-1)(r-2) & (r-1)(k-r+1) \\ (r-1)(k-r+1) & (k-r)(k-r+1) \end{pmatrix},$$

$$(3.3.14) \qquad (p^2_{jk}) = \begin{pmatrix} r(r-1) & r(k-r) \\ r(k-r) & (k-r)^2+(r-2) \end{pmatrix}.$$

(e) The negative Latin square association scheme. This important
class was discovered by Mesner (1967) and is defined by the following
parameters

$$(3.3.15) \qquad v = k^2, \quad n_1 = r(k+1), \quad n_2 = (k-r-1)(k+1),$$

R. C. BOSE

$$(3.3.16) \quad (p_{jk}^1) = \begin{pmatrix} (r+1)(r+2)-(k+2) & (k-r-1)(k+1) \\ (k-r-1)(k+1) & (k-r-1)(k-r) \end{pmatrix},$$

$$(3.3.17) \quad (p_{jk}^2) = \begin{pmatrix} r(r+1) & r(k-r) \\ r(k-r) & (k-r)(k-r+1)-k-2 \end{pmatrix}.$$

Examples of this association scheme will occur in subsequent chapters.

4. **Association matrices.** We define

$$(3.4.1) \quad B_i = (b_{\alpha i}^\beta) = \begin{bmatrix} b_{1i}^1 & b_{1i}^2 & \cdots & b_{1i}^v \\ \cdots\cdots\cdots\cdots\cdots \\ b_{vi}^1 & b_{vi}^2 & \cdots & b_{vi}^v \end{bmatrix}$$

where

$b_{\alpha i}^\beta = 1$, if the objects α and β are ith associates

$= 0$, otherwise.

B_i is a symmetric matrix, in which each row total and each column total is n_i.

Among the numbers

$$b_{\alpha 0}^\beta, \quad b_{\alpha 1}^\beta, \quad \cdots, \quad b_{\alpha m}^\beta$$

only one is unity, i.e., $b_{\alpha i}^\beta$ if α and β are i-th associates. Hence

$$(3.4.2) \quad B_0 + B_1 + \cdots + B_m = J_v,$$

where J_v is the $v \times v$ matrix each of whose elements is unity.

It also follows that the linear form

$$(3.4.3) \quad c_0 B_0 + c_1 B_1 + \cdots + c_m B_m$$

is equal to the zero matrix if and only if

$$c_0 = c_1 = \cdots = c_m = 0;$$

hence the linear functions of B_0, B_1, $\cdots$, B_m form a vector space with basis B_0, B_1, $\cdots$, B_m. One can now prove [Bose and Mesner (1959)],

Lemma (3.4.1).

$$(3.4.4) \qquad \sum_{\gamma=1}^{v} b_{\alpha j}^{\gamma} b_{\gamma k}^{\beta} = p_{jk}^{0} b_{\alpha 0}^{\beta} + \cdots + p_{jk}^{i} b_{\alpha i}^{\beta} + \cdots + p_{jk}^{m} b_{\alpha m}^{\beta}.$$

We now note that the left-hand side of (3.5.4) is the element in the αth row and βth column of the product $B_j B_k$, and $b_{\alpha i}^{\beta}$ is the element in the αth row and βth column of B_i ($i = 0$, $\cdots$, m). Thus

$$(3.4.5) \qquad B_j B_k = p_{jk}^{0} B_0 + p_{jk}^{1} B_1 + \cdots + p_{jk}^{m} B_m.$$

The product of two matrices of the form (3.4.3), where the c_i are scalars, may be expressed as a linear combination of terms of the form $B_j B_k$ and will reduce to the form (3.4.3). The set of matrices of this form is therefore closed under multiplication. It is clear that it forms an Abelian group under addition. Thus the linear functions of B_0, B_1, $\cdots$, B_m form a ring with unit element, which will be a linear associative algebra if the coefficients c_i range over a field. Multiplication is also commutative.

By evaluating $B_i B_j B_k$ in two different ways using (3.4.5) it follows as a consequence of the associative low of multiplication that

$$(3.4.6) \qquad \sum_u p_{ij}^{u} p_{uk}^{t} = \sum_u p_{jk}^{u} p_{iu}^{t}.$$

In these equations the summation over u runs from 0 to m and the remaining indices are arbitrary but fixed,

$$0 \leq i, j, k, t \leq m$$

Now let us define P_k by

$$P_k = (p_{ik}^j) = \begin{bmatrix} p_{0k}^0 & p_{0k}^1 & \cdots & p_{0k}^m \\ p_{1k}^0 & p_{1k}^1 & \cdots & p_{1k}^m \\ \cdots\cdots\cdots\cdots\cdots \\ p_{mk}^0 & p_{mk}^1 & & p_{mk}^m \end{bmatrix}, \quad k = 1, 0, \cdots, m .$$

Now the left side of (3.4.6) is the element in the i-th row and t-th column of $P_j P_k$. Also the element in the i-th row and t-th column of P_u is p_{iu}^t, so that the right side of (3.5.6) is the element in the i-th row and t-th column of

$$p_{jk}^0 P_0 + p_{jk}^1 P_1 + \cdots + p_{jk}^m P_m .$$

Hence we have

(3.4.7) $$P_j P_k = p_{jk}^0 P_0 + p_{jk}^1 P_1 + \cdots + p_{jk}^m P_m.$$

Thus, the P's multiply in the same manner as the B's. Since $p_{0k}^i = 1$ if $k = i$ and 0 otherwise, the 0th row of P_k contains a 1 in column k and 0's in other positions, which is enough to show that if

$$c_0 P_0 + c_1 P_1 + \cdots + c_m P_m = 0 ,$$

then

$$c_0 = c_1 = \cdots = c_m = 0 ;$$

i.e., P_0, P_1, $\cdots$, P_m are linearly independent. They thus form the basis for a vector space and combine in the same way as the B's under addition, as well as under multiplication. They provide a regular representation in

$$(m + 1) \times (m + 1)$$

matrices of the algebra given by the B's, which are $v \times v$ matrices. In particular, $P_0 = I_{m+1}$.

Since the B's are commutative, the P's are commutative. In general they are not incidence matrices and are not symmetric. P_k does not have equal row totals, but has the same equal column totals n_k as B_k. In analogy with (3.4.2), all elements of row j of $\sum_k P_k$ are equal to n_j. Let

$$B = c_0 B_0 + c_1 B_1 + \cdots + c_m B_m$$

be any element of our algebra, and let $f(\lambda)$ be a polynomial. Then we can express

$$f(B) = \ell_0 B_0 + \ell_1 B_1 + \cdots + \ell_m B_m.$$

If

$$P = c_0 P_0 + c_1 P_1 + \cdots + c_m P_m$$

is the representation of B, then

$$f(P) = \ell_0 P_0 + \ell_1 P_1 + \cdots + \ell_m P_m.$$

Let $f(\lambda)$ be the minimum function of B and $\phi(\lambda)$ the minimum function of P. Then $f(\lambda)$ is the monic polynomial of least degree for which

$$f(B) = 0.$$

$$f(B) = 0 \rightarrow \ell_0 = \ell_1 = \cdots = \ell_m \rightarrow f(P) = 0;$$

i.e., $f(\lambda)$ is divisible by $\phi(\lambda)$.

Similarly $\phi(\lambda)$ is divisible by $f(\lambda)$. Since both are monic polynomials,

$$f(\lambda) = \phi(\lambda).$$

That is, B and P have the same distinct characteristic roots, and every matrix B has at most $m+1$ distinct characteristic roots, which are solutions of the minimum equation of P.

R. C. BOSE

5. Combinatorial applications of the algebra of association matrices.

(a) Consider a PBIB design based on an m class association scheme, with the association matrices B_i defined by (3.4.1). Let

(3.5.1) $N = (n_{ij})$, $i = 1, 2, \ldots v$, $j = 1, 2, \ldots, b$;

be the indicence matrix of the design, i.e. $n_{ij} = 1$ or 0 according as the treatment i does or does not occur in the j-th block. Then

(3.5.2) $B = NN' = rB_0 + \lambda_1 B_1 + \cdots + \lambda_m B_m$,

(3.5.3) $P = rP_0 + \lambda_1 P_1 + \cdots + \lambda_m P_m$.

The elements of NN' are non-negative and for connected designs, i.e. designs in which every treatment contrast is estimable, NN' is irreducible. Also in virtue of the identity (3.2.9) the sum of the elements in every row of NN' is rk. Hence

(3.5.4) $B^* = \dfrac{1}{rk} B = \dfrac{1}{rk} NN'$,

is a stochastic matrix (i.e. an irreducible matrix for which the sum of each row is unity). For such a matrix [Brauer (1952)] unity is a simple root and is greater than all the other roots. Hence rk is a simple root of B and is, therefore, also a simple root of P.

One can now show [Bose (1963 b)] that the m characteristic roots of P other than rk are the roots of the matrix.

(3.5.5) $P^* = (p^*_{ij})$ $i, j = 1, 2, \ldots, m$;

where

(3.5.6) $p^*_{ij} = r\delta_{ij} + \lambda_1 p^j_{i1} + \cdots + \lambda_m p^j_{im} - n_i \lambda_i$

δ_{ij} being the Kronecker delta.

(b) For a balanced incomplete block (BIB) design we have derived in

Chapter I, the inequality $b \geqq v$ due to Fisher (1940), where b is the number of blocks and v is the number of treatments. One may ask what the corresponding result is for PBIB designs. Now

$$b \geqq \text{rank } N$$

$$\geqq \text{rank } NN'.$$

But the rank of NN' is v, unless NN' is singular, i.e., has a zero characteristic root, in which case P and therefore P^* has a zero characteristic root. Thus:

A necessary condition for $b \leqq v$ in a PBIB design is

(3.5.7) $$|p_{ij}^*| = 0$$

where p_{ij}^* is given by (3.5.6).

Thus Fisher's inequality $b > v$ is satisfied in general. It can be violated by only those designs for which (3.5.7) is satisfied. This result is due to Nair (1943). An alternative proof will be found in Bose (1952).

Bose and Mesner (1959) have given a general method of calculating the multiplicaties of the roots of $B = NN'$. We shall illustrate the determination of the multiplicaties $\alpha_0 = 1$, α_1, α_2 of the roots $\theta_0 = rk$, θ_1, θ_2, in the special case m = 2. Now θ_1 and θ_2 are the characteristic roots of P^* given by (3.5.5). Setting

(3.5.8) $$\gamma = p_{12}^2 - p_{12}^1, \quad \beta = p_{12}^1 + p_{12}^2, \quad \Delta = \gamma^2 + 2\beta + 1,$$

we find after some calculation that

$$\theta_1 = r + \tfrac{1}{2}[(\lambda_1 - \lambda_2)(\sqrt{\Delta} + \gamma) - (\lambda_1 + \lambda_2)],$$

$$\theta_2 = r - \tfrac{1}{2}[(\lambda_1 - \lambda_2)(\sqrt{\Delta} - \gamma) + (\lambda_1 + \lambda_2)].$$

Now

(3.5.9)
$$\text{tr } I = 1 + \alpha_1 + \alpha_2 = v$$

(3.5.10)
$$\text{tr } NN' = rk + \alpha_1\theta_1 + \alpha_2\theta_2 = vr,$$

whence after some calculation [Connor and Clatworthy (1954)]

(3.5.11)
$$\alpha_1 = \frac{n_1 + n_2}{2} - \frac{(n_1 - n_2) + \gamma(n_1 + n_2)}{2\sqrt{\Delta}},$$

(3.5.12)
$$\alpha_2 = \frac{n_1 + n_2}{2} + \frac{(n_1 - n_2) + \gamma(n_1 + n_2)}{2\sqrt{\Delta}}$$

It is interesting to note that the multiplicities α_1 and α_2 depend only on the parameters of the association scheme, i.e. parameters of the first kind. The corresponding general result is due to Bose and Mesner (1959).

Since the multiplicities α_i are expressible in terms of the parameters of the association scheme, we cannot have a set of parameters leading to nonintegral values α_i. This fact can be used to prove the impossibility of certain association schemes.

We are now in a position to see how Fisher's inequaulity should be modified for the case when P and therefore $P*$ has a zero characteristic root. Let α be the multiplicity of the root zero of $B = NN'$. Then

$$b \geq \text{rank } N$$

$$\geq \text{rank } NN'$$

$$= v - \alpha.$$

Hence Fisher's inequality is replaced by [Connor and Clatworthy (1954)]

(3.5.13)
$$b \geq v - \alpha.$$

6. **Final remarks.** For the construction and further properties of BIB designs the reader may refer to the following papers: Archbold and

R. C. BOSE

Johnson (1956), Bose and Nair (1939), Bose and Shimamoto (1952), Bose, Clatworthy and Shrikhande (1954), Clatworthy [(1954), (1955), (1956)], Masuyama [(1961), (1964 a), (1964 b)], Nair [(1950, (1951 a), (1951 b)], Ogawa [(1959), (1960)], Ray-Chaudhuri [(1962 b), (1965)] and Shrikhande (1965).

CHAPTER IV

STRONGLY REGULAR GRAPHS AND PARTIAL GEOMETRIES

1. **Strongly regular graphs.** A finite graph G consists of a finite set of v vertices, and a relation adjacency such that any two distinct vertices of G may be either adjacent or non-adjacent. Adjacent vertices may be said to be joined and non-adjacent vertices to be unjoined. We shall be concerned with finite graphs only, and use the word graph in the sense of finite graphs.

The graph G is said to be regular (of valence n_1) if each vertex of G is joined to exactly n_1 other vertices. In this case each vertex will be unjoined to exactly n_2 other vertices, where

(4.1.1) $$n_1 + n_2 = v - 1.$$

A regular graph G will be said to be strongly regular if (i) any two vertices which are joined in G, are both simultaneously joined to exactly p_{11}^1 other vertices (ii) any two vertices which are unjoined in G, are both simultaneously joined to exactly p_{11}^2 vertices.

A strongly regular graph G thus depends on four parameters

(4.1.2) $$v, n_1, p_{11}^1, p_{11}^2$$

where n_2 is given by (4.1.1). The concept of a strongly regular graph is isomorphic to that of a 2-class association scheme. The v vertices correspond to the v objects or treatments of the association scheme and two distinct vertices of G are adjacent or non-adjacent according as the corresponding treatments of the association scheme are first associates or second associates. Note that the constancy of v, n_1, p_{11}^1 and p_{11}^2 guarantees the constancy of all the other parameters of the association scheme [Bose and Clatworthy (1955)] . Thus it will be convenient to call two vertices of a strongly regular graph G first associates if they are adjacent and second associates if they are non-adjacent. Then given any two vertices x and y which are i-th associates, the number of vertices which are simultaneously j-th associates of x and k-th associates of y is $p_{jk}^i [i, j, k = 1, 2]$. Also a vertex may be called its own 0-th associate. Hence we may allow i, j, k to take also the value 0. The formulae (3.2.4) through (3.2.9) and which were proved for association schemes remain valid. In particular we have

(4.1.3) $$p_{12}^1 = n_1 - p_{11}^1 - 1 = p_{21}^1, \quad p_{22}^1 = n_2 - n_1 + p_{11}^1 + 1$$

(4.1.4) $$p_{12}^2 = n_1 - p_{11}^2 = p_{21}^2, \quad p_{22}^2 = n_2 - n_1 + p_{11}^2 - 1$$

(4.1.5) $$n_1 p_{12}^1 = n_2 p_{11}^2, \quad n_1 p_{22}^1 = n_2 p_{12}^2.$$

The adjacency matrix of a graph G with v vertices is defined to be the v × v matrix $A = (a_{ij})$ where $a_{ij} = 1$ if the i-th and j-th vertices are adjacent and zero otherwise. In particular $a_{ii} = 0$. When G is strongly regular with parameters (v, n_1, p_{11}^1, p_{11}^2) and B_0, B_1, B_2 are the association matrices of the corresponding association scheme then B_1 is the ad-

jacency matrix of G. The complementary graph $\overline{G}$ of a graph G is defined
to be a graph with the same vertices as G but with the relation of adjacency and non-adjacency reversed. When G is strongly regular then the
adjacency matrix of $\overline{G}$ is B_2. We can now define the matrices of P_0, P_1,
P_2 as in Chatper III and all the formulae regarding the characteristic
roots or eigenvalue of the linear functions of P_0, P_1, P_2 or B_0, B_1, B_2
can be taken over.

2. <u>Seidel equivalence of strongly regular graphs</u>. Let G be a
strongly regular graph with parameters

$$v, \; n_1, \; p_{11}^1, \; p_{11}^2.$$

We can obtain another graph G* from it by the following process:
Let the set of vertices V of G be divided into disjoint subsets, V_1 and
V_2, $V = V_1 \cup V_2$. G* has the same set of vertices as G. Two vertices of
G* both of which belong to V_1 or to V_2 are adjacent or non-adjacent in G*
according as they are adjacent or non-adjacent in G. Two vertices of G*
one of which belongs to V_1 and the other to V_2 are adjacent in G* if they
are non-adjacent in G, and non-adjacent in G* if they are adjacent in G.
Then G* may be said to be derived from G by complementation with respect
to V_1 and V_2. If G* is strongly regular it is defined to be Seidel
equivalent to G, or more briefly S-equivalent to G [Seidel (1967)].

Let $|V_1| = v_1$, $|V_2| = v_2$, then $v = v_1 + v_2$. In writing down the adja-
cency matrix of G, we may take the first v_1 rows (columns) to correspond
to the vertices in V_1 and the last v_2 rows (columns) to correspond to the
vertices in V_2. Then we can write the adjacency matrix of G as

(4.2.1)
$$A = \begin{bmatrix} A_{11} & A_{12} \\ A_{21} & A_{22} \end{bmatrix}$$

where A_{11} and A_{22} are square matrices of order v_1 and v_2 respectively, A_{12} is a $v_1 \times v_2$ matrix, and $A_{21} = A_{12}'$. Then clearly the adjacency matrix of G^* is

$$(4.2.2) \qquad A^* = \begin{bmatrix} A_{11} & \overline{A}_{12} \\ \overline{A}_{21} & A_{22} \end{bmatrix}.$$

One can now investigate the conditions under which G^* is strongly regular and therefore by definition is S-equivalent to G. In this connection Bose and Shrikhande (1970) proved the following theorems:

Theorem (4.2.1). Let G be a strongly regular graph with parameters

$$v, \ n_1, \ p_{11}^1, \ p_{11}^2.$$

If the vertices of G are divided into two disjoint subsets V_1 and V_2, where $|V_1| = v_1$, $|V_2| = v_2$, then the necessary and sufficient conditions for the graph G^* derived from G by complementation with respect to V_1, V_2, to be strongly regular are

(a) In G each vertex in V_1 is adjacent to w_1 vertices in V_1 (and therefore $n_1 - w_1$ vertices in V_2); also each vertex in V_2 is adjacent to w_2 vertcies in V_2 (and therefore to $n_1 - w_2$ vertices in V_1), where

$$w_1 - w_2 = \frac{v_1 - v_2}{2}$$

(b) $$p_{11}^1 + p_{11}^2 = 2n_1 - \frac{v}{2}.$$

When these conditions are satisfied the parameters of G^* are given by

$$(4.2.3) \qquad v^* = v, \ n_1^* = 2w_1 + v_2 - n_1 = 2w_2 + v_1 - n_1,$$

$$(4.2.4) \qquad p_{11}^{1*} = n_1^* - n_1 + p_{11}^1, \ p_{11}^{2*} = n^* - n_1 + p_{11}^2.$$

If the graph G^* is required to have the same parameters as G, then

$n_1^* = n_1$. This automatically ensures that $p_{11}^{1*} = p_{11}^1$ and $p_{11}^{2*} = p_{11}^2$. Also $n_1 - w_1 = \frac{v_2}{2}$, $n_1 - w_2 = \frac{v_1}{2}$, i.e., in G each vertex of V_1 is adjacent to exactly half the vertices in V_2, and each vertex in V_2 is adjacent to exactly half the vertices in V_1. We therefore have

Theorem (4.2.2). Let G be a strongly regular graph with parameters

$$v, \; n_1, \; p_{11}^1, \; p_{11}^2.$$

If the vertices of G are divided into two disjoint subsets V_1 and V_2, then the necessary and sufficient conditions for the graph G* derived from G by complimentation with respect to V_1 and V_2, to be strongly regular with the same parameters as G are

(a) In G each vertex in V_1 is adjacent to exactly half the vertices in V_2, and each vertex in V_2 is adjacent to exactly half the vertices in V_1.

(b) $$p_{11}^1 + p_{11}^2 = 2n_1 - \frac{v}{2}.$$

3. **Partial geometries and the corresponding PBIB designs.** A partail geometry (r, k, t) is a system of points and lines, and a relation of incidence between them satisfying the following axioms:

A1. Any two distinct points are incident with not more than one line.

A2. Each point is incident with r lines.

A3. Each line is incident with k points.

A4. If the point P is not incident with the line ℓ, there are exactly t lines (t $\geq$ 1) which are incident with P, and also incident with some point incident with ℓ.

Clearly $1 \leq t \leq k$, $1 \leq t \leq r$.

(a) If there were two distinct lines ℓ and m each incident with two distinct points P_1 and P_2, then A1 would be contradicted. Hence:

A'1. Any two distinct lines are incident with not more than one point.

Given a partial geometry (r, k, t), there exists a dual partial geometry (k, r, t), obtained by calling the points of the first, the lines of the second; and the lines of the second the points of the first.

The above follows by noting the duality of A1 and A'1, the duality of A2 and A3, and the self-dual nature of A4.

For convenience we may introduce the ordinary geometric language. Thus if a point is incident with a line we say that the point lies on the line, or is contained in the line, and the line passes through the point. If two points are incident on a line we speak of the line as joining the two points. If a point is incident with each of two lines, we say that the lines intersect in that point. With this language A4 may be re-phrased as:

A4. Through any point P not lying on a line ℓ, there pass exactly t lines intersecting ℓ.

4. **Graph of a partial geometry**. The graph G of a partial geometry (r, k, t) is defined as follows. The vertices of G are the points of the partial geometry. Two vertices of G are joined (adjacent) if the corres-ponding points of the geometry are joined (incident with the same line). Two vertices of G are unjoined (non-adjacent) if the corresponding points of the partial geometry are unjoined (i.e. there exists no line incident with both the points).

Theorem (4.4.1). The graph of partial geometry (r, k, t) is strong-ly regular with parameters

R. C. BOSE

$$v = k[(r-1)(k-1) + t]/t, \quad n_1 = r(k-1)$$

$$p_{11}^1 = (t-1)(r-1) + k-2, \quad p_{11}^2 = rt,$$

$$1 \le t \le r, \quad 1 \le t \le k.$$

Let there be v points and b lines in the partial geometry. Since the points of the geometry have been identified with the vertices of the graph G, we can call two points of the geometry first associates if they are joined by a line, and second associates if they are not joined by a line. Now through any point P of the geometry there pass r lines, each of which contains $k-1$ other points besides P. Hence P has exactly $r(k-1)$ first associates. Hence

(4.4.1) $$n_1 = r(k-1).$$

This shows that G is a regular graph. Consider the $b-r$ lines not passing through P. From A4 each of these lines contains exactly t first associates of P. Any particular first associate Q of P, lies on $r-1$ such lines, since one of the r lines passing through Q joins it to P. Hence the number of first associates is

(4.4.2) $$n_1 = t(b-r)/(r-1).$$

By similar arguments we can prove [Bose (1963 b)] that

(4.4.3) $$n_2 = (k-t)(b-r)/r$$

(4.4.4) $$p_{11}^1 = (t-1)(r-1) + (k-2), \quad p_{11}^2 = rt$$

Comparing (4.4.1) and (4.4.2) we have

(4.4.5) $$b = r[(r-1)(k-1) + t]/t,$$

and substituting for b in (4.4.3) we have

R. C. BOSE

(4.4.6) $n_2 = (r-1)(k-1)(k-t)/t$

We have now verified that G is strongly regular. The value of v is given by (4.1.1) and the other parameters are obtained from (4.1.3) and (4.1.4). Hence the parameters of the strongly regular graph (or the assoc- iation scheme) corresponding to a partial geometry (r, k, t) are given by

(4.4.7)

$$v = k[(r-1)(k-1)+t]/t, \quad n_1 = r(k-1), \quad n_2 = (r-1)(k-1)(k-t)/t,$$

(4.4.8) $(p_{jk}^1) = \begin{pmatrix} (t-1)(r-1)+k-2 & (r-1)(k-t) \\ (r-1)(k-t) & (r-1)(k-t)(k-t-1)/t \end{pmatrix}$,

(4.4.9) $(p_{jk}^2) = \begin{pmatrix} rt & r(k-t-1) \\ r(k-t-1) & \dfrac{(r-1)(k-1)(k-t)}{t}-r(k-t-1)-1 \end{pmatrix}$.

This association scheme may be called the geometric association scheme with characteristics (r, k, t).

Corollary. For a partial geometry (r, k, t) the number v of points is given by (4.4.7), and the number b of lines is given by (4.4.5).

If the v points of the partial geometry (r, k, t) are taken as the treatments, and the k lines are taken as the blocks of a design then it is clear that a partial geometry (r, k, t) is a PBIB design based on the association scheme with parameters (4.4.7), (4.4.8) and (4.4.9), for which the parameters of the second kind are

(4.4.10) $r, k, \lambda_1 = 1, \lambda_2 = 0$.

Bose and Clatworthy (1955) considered two class PBIB designs with $r < k$, $\lambda_1 = 1$, $\lambda_2 = 0$. From the results of Chapter III, paragraph 5, it follows that for such designs the matrix P^* given by (3.5.5) has a zero characteristic root, if we take $\lambda_1 = 1$, $\lambda_2 = 0$. Hence

R. C. BOSE

$$rp_{12}^1 - (r-1)p_{12}^2 = r(r-1).$$

Using this relation, and the identities (3.1.3), (3.1.4), (3.2.1), (3.2.2), (3.2.3), (3.2.4) they showed that the parameters of the design must be given by (4.4.7), (4.4.8), (4.4.9). This raises the interesting question, whether the design is a partial geometry. The answer is in the affirmative. Since the axioms A1, A2, A3 are evidently satisfied, it only remains to show that A4 is also satisfied.

Let K be the st of k treatments contained in a particular block, and let $\overline{K}$ be the set of remaining $v - k$ treatments. Let $g(x)$ denote the number of treatments in $\overline{K}$ which have exactly x first associates in K. Then easy counting arguments show that

(4.4.11) $$\sum_{x=0}^{k} g(x) = v - k = k(k-1)(r-1)/t.$$

(4.4.12) $$\sum_{x=0}^{k} x\, g(x) = k(n_1 - k + 1) = k(r-1)(k-1).$$

(4.4.13) $$\sum_{x=0}^{k} x(x-1)g(x) = k(k-1)(p_{11}^1 - k + 2) = k(k-1)(t-1)(r-1).$$

Hence $\overline{x}$, the average value of x, is

$$\overline{x} = \sum xg(x)/\sum g(x) = t,$$

and

(4.4.14) $$\sum_{x=0}^{k} g(x)\,(x-\overline{x})^2 = \sum_{x=0}^{k} g(x)\,(x-t)^2 = 0.$$

Hence x must always have the value t. This is equivalent to the axiom A4. Hence a PBIB desing with r replications, block size k, $\lambda_1 = 1$, $\lambda_2 = 0$, is

R. C. BOSE

a partial geometry (r, k, t) if r < k.

One may ask whether a partial geometry (r, k, t) exists for all values of r, k, t. Now for the corresponding PBIB design the multiplicity α_1 of the characteristic root θ_1 of the incidence matrix NN' is given by (3.5.11). Substituting for n_1, n_2, v and Δ from (4.4.7), (4.4.8), (4.4.9) and (3.5.8) we have

(4.4.15)
$$\alpha_1 = \frac{rk(r-1)(k-1)}{t(k+r-t-1)} .$$

Hence a necessary condition for the existence of a partial geometry (r, k, t) is that the number α_1 given by (4.4.15) is a positive integer. For example if r = 3, t = 1 then the only possible values of k are k = 2, 3, 5 and 11. The cases k = 2, 3, 5 are possible, but a rather lengthy combinatorial argument [Bose and Clatworthy (1955)] shows the case k = 11 to be impossible.

CHAPTER V

THE FUNDAMENTAL CHARACTERIZATION THEOREM

1. <u>Weakly balanced designs and graphs</u>. Let G be a finite regular graph of valence d. Let $\Delta(x,y)$ denote the number of vertices which are simultaneouly adjacent to two given distinct vertices x and y. Then G is defined to be edge regular of edge-degree δ if $\Delta(x,y) = \delta$ for every pair x, y of adjacent vertices. We shall consider here a particular class of edge regular graphs.

Consider a set of v objects or treatments arranged into b blocks or sets, such that each block contains at least two treatments, and the treatments in a given block are all distinct. Two treatments θ and ϕ are called first associates if there is a block containing both θ and ϕ. Otherwise they are second associates. The arrangement is called a weakly balanced design if the following conditions hold

(1) Each treatment occurs in r blocks,

(2) Two distinct treatments do not occur in more than one block,

(3) Each treatment has exactly $r(k-1)$ first associates,

(4) Given any two treatments θ and ϕ which are first associates there are exactly δ treatments which are first associates to both θ and ϕ.

The condition (3) means that the sum of the sizes of the blocks in which a given treatment θ appears is rk. Hence k is the average size of the blocks in which a given treatment appears. Since each block contains at least two treatments $k \geqslant 2$.

For any treatment θ there exists at least one block B containing θ for which the block size $k_1 \geq k$. Let ϕ be any other treatment in B. Then the δ treatments which are by (4) first associates to both θ and ϕ are made up of the $k_1 - 2$ treatments in B other than θ and ϕ, together with treatments which occur simultaneouly in a block containing θ but not ϕ, and a block containing ϕ and not θ. If there are n such treatments then $\delta = k_1 - 2 + n \geqslant k - 2$. Let

(5.1.1) $\delta = k - 2 + \alpha$

Then $\alpha \geqslant 0$.

For any pair of treatments θ and ϕ, let $\Delta(\theta,\phi)$ be the number of treatments which are first associates of both θ and ϕ. Then $\Delta(\theta,\phi) = \delta$ if θ and ϕ are first associates. Let β be the upper bound $\Delta(\theta,\phi)$ for all

R. C. BOSE

pairs θ, ϕ such that θ and ϕ are second associates. Then β is a non-negative integer.

The weakly balanced design under consideration will be said to have the parameters (r, k, α, β). Note that a partial geometry (r, k, t) is a weakly balanced design (r, k, α, β) with

$$(5.1.2) \qquad \alpha = (r-1)(t-1), \quad \beta = rt \,,$$

the blocks and treatments of the design being the lines and points of the partial geometry.

The graph of a weakly balance (WB) design is defined to be a graph whose vertices correspond to the treatments of a design and for which two vertices are adjacent if and only if the corresponding treatments are first associates. The graph of a WB-design is called a WB-graph. Clearly the graph of a WB design with parameters (r, k, α, β) satisfies the following conditions

    (c_1) G is regular of valence $d = r(k-1)$,

    (c_2) G is edge-regular with edge-degree $\delta = k - 2 + \alpha$,

    (c_3) $\Delta(x,y) < \beta$, for all pairs of non-adjacent vertices, x and y of G.

Here d, δ, r are positive integers, $\beta \geqslant 0$ is a non-negative integer and $k \geqq 2$, $\alpha > 0$.

A finite graph G satisfying the above conditions will be called a pseudo WB-graph. A pseudo WB-graph may not necessarily be the graph of a WB-design.

We shall show that if k is sufficiently large in comparison to r, α and β, then a pseudo WB-graph will be the graph of a WB-design.

    2. <u>Definitions</u>. We define here some functions of the parameters

R. C. BOSE

r, k, α, β which play an important part in the investigations which follow.

(5.2.1) $\qquad \gamma(r,\alpha) = 1 + (r-1)\alpha$

(5.2.2) $\qquad q(r,\alpha) = 1 + (2r-1)\alpha$

(5.2.3) $\qquad \rho(r,\alpha,\beta) = \beta + (2r-1)\alpha$

(5.2.4) $\qquad p(r,\alpha,\beta) = 1 + \frac{1}{2}(r+1)(r\beta - r - 2\alpha)$

We shall denote as usual the cardinality of a set S by $|S|$.

A clique K of a graph is a set of vertices adjacent to each other.
A clique K will be called complete if we cannot find a vertex x, not contained in K such that xUK is a clique. Thus a complete clique cannot be
extended to a larger clique by the adjunction of a new vertex belonging
to the graph.

Now consider the graph G with the properites, (c_1), (c_2), (c_3). A
clique K of G will be called a major clique if

(5.2.5) $\qquad |K| \geq 1 + k - \gamma(r,\alpha) = k - (r-1)\alpha$.

The clique K of G will be called a grand clique if it is both major
and complete.

A claw [p,S] of G, consists of a vertex p, the vertex of the claw,
and a non-empty set S of vertices of G, not containing p, such that p is
adjacent to every vertex in S, but any two vertices in S are non-adjacent.
The order of the claw is defined to be the number $s = |S|$.

. 3. Theorems and lemmas for claws in pseudo WB-graphs. Let G be a
pseudo WB-graph satisfying conditions (5.1.3).

Theorem (5.3.1) If $k > p(r,\alpha,\beta)$, there cannot exist a claw of order
$r+1$ in G.

R. C. BOSE

Suppose there exists in G a claw $[p,S]$ of order s. Let T be the set of vertices of G, not belonging to $[p,S]$, and adjacent to p. Let $f(x)$ denote the number of vertices q in T, such that q is adjacent to exactly x vertices in S. Counting the number of vertices in T we have from (c_1),

$$(5.3.1) \qquad \sum_{x=0}^{s} f(x) = d - s .$$

Counting the number of ordered pairs (b,q) where b and q are adjacent, b belongs to S, and q belongs to T, we have from (c_2)

$$(5.3.2) \qquad \sum_{x=0}^{s} xf(x) = \delta s .$$

Again counting the triplets (b_1, b_2, q), where b_1, b_2 is an ordered pair of vertices in S, q is a vertex in T, and $b_1 b_2$, are both adjacent to q, we have from (c_3)

$$(5.3.3) \qquad \sum_{x=0}^{s} x(x-1)f(x) \leqq s(s-1)(\beta-1).$$

If a claw of order $r+1$ exists, putting $s = r+1$, we have from $(5.3.1)$, $(5.3.2)$ and $(5.3.3)$:

$$(5.3.4) \quad f(0) + \frac{1}{2} \sum_{x=1}^{r+1} (x-1)(x-2)f(x) \leqq d - s(1+\delta) + \frac{1}{2}s(s-1)(\beta-1)$$

$$= -k + 1 + \frac{1}{2} (r+1)(r\beta - r - 2\alpha)$$

$$= -k + p(r,\alpha,\beta) .$$

Since the left-hand side is essentially non-negative whereas $k > p(r,\alpha,\beta)$ by hypothesis, we have a contradiction. This proves our theorem.

The following Lemmas are readily proved:

Lemma $(5.3.1)$. If $k > \gamma(r,\alpha)$, then any claw of G of order $s < r$, can be extended to a claw of order r.

R. C. BOSE

Lemma (5.3.2). Given a claw $[p,S]$ of G of order $r-1$ there exist at least $k-\gamma(r,\alpha)$ distinct vertices q of G such that $[p,SUq]$ is a claw of order r.

4. **Theorems and lemmas for cliques in pseudo WB-graphs.** As before let G be a graph satisfying the conditions (c_1), (c_2), (c_3).

Lemma (5.4.1). If $k > \max[\gamma(r,\alpha), p(r,\alpha,\beta)]$, then any pair of adjacent vertices p and q is contained in at least on grand clique.

From Lemma (5.3.1) we can extend the claw $[p,q]$ to a claw $[p,S]$ of order r. Let b_1, b_2, ..., b_r be vertices in S other than q. Let Ω be the set of vertices ω, which when adjoined to $S-q$ give a claw $[p,S^*]$ of order r, where $S^* = (S-q)U\omega$. Of course q is contained in Ω and from Lemma (5.3.2)

$$|\Omega| \geqq k - \gamma(r,\alpha).$$

The vertices in Ω are all adjacent to one another. If any two were not adjacent they could be added to b_1, b_2, ..., b_r to give a claw of order $r+1$, which would contradict Theorem (5.3.1). Let $K = pU\Omega$. Then K is a major clique since

$$|K| \geq 1 + k - \gamma(r,\alpha).$$

We can extend the major clique K by adding new vertices till it is complete and therefore a grand clique.

The following Lemmas are easily proved:

Lemma (5.4.2). If K and L are cliques of G, and KUL is not a clique, then $|K \cap L| \leqq \beta$.

Lemma (5.4.3). If K and L are cliques of G and $K \cap L$ contains at least two vertices a and b, then $|KUL| \leq k + \alpha$.

Lemmas (5.4.4). If K and L are cliques of G, KUL is not a clique and $K \cap L$ contains at least two vertices, then

$$|K| + |L| \leqslant k + \alpha + \beta .$$

Theorem (5.4.1). If $k > \max[\gamma(r,\alpha), \rho(r,\alpha,\beta), p(r,\alpha,\beta)]$ then any pair of adjacent vertices p and q is contained in one and only one grand clique.

The existence of at least one grand clique follows at once from Lemma (5.4.1).

Suppose there exist at least two distinct grand cliques K and L both containing the adjacent vertices p and q. Since K and L are complete, $K \cup L$ is not a clique. Hence from Lemma (5.4.4)

$$|K| + |L| \leqq k + \alpha + \beta .$$

But K and L are both major cliques. Hence

$$|K| + |L| \geqq 2\{k - (r-1)\alpha\} ,$$

which shows that

$$k \leqq \beta + (2r - 1)\alpha = \rho(r,\alpha,\beta) ,$$

contrary to the hypothesis.

Theorem (5.4.2). If $k > \max[\gamma(r,\alpha), \rho(r,\alpha,\beta), p(r,\alpha,\beta)]$ then each vertex of G is contained in exactly r grand cliques.

From Theorem (5.4.1) any pair of adjacent vertices is contained in exactly one grand clique. Also from Lemma (5.3.1), p is the vertex of at least one claw of order r. Let[p, S] be a claw of order r, where $S = \{b_1, b_2, \ldots, b_r\}$. As in Theorem (5.3.1), let T be the set of vertices not belonging to S, which are adjacent to p.

Let H_j be the set consisting of p, b_j, and q belonging to T, such that q is adjacent to b_j but not adjacent to b_i, $i \neq j$. As in Theorem

(5.3.1) let $f(x)$ denote the number of vertices in T which are adjacent to exactly x vertices in S. Then $f(0) = 0$, otherwise there would exist a claw of order $r+1$. Putting $s = r$ in (5.3.1) and (5.3.2), we have

$$(5.4.1) \qquad \sum_{x=1}^{r} f(x) = r(k-2).$$

$$(5.4.2) \qquad \sum_{x=1}^{r} xf(x) = r(k-2+\alpha).$$

$$(5.4.3) \qquad f(1) \geqslant r(k-2-\alpha).$$

Any two vertices of H_j are adjacent to one another, otherwise there would exist a claw of order $r+1$. Thus H_j is a clique.

Put $H_j^* = H_j - (b_j \cup p)$. Then H_j^* consists of exactly those vertices of T which are adjacent to b_j but to no other vertex of S. Hence H_1^*, H_2^*, ..., H_r^* are disjoint sets, and the total number of vertices in these sets is $f(1)$.

Now there is a unique grand clique K_j containing b_j and p. The number of vertices in K_j cannot be less than the number of vertices in H_j. If possible let $|K_j| < |H_j|$. Since K_j is a grand clique it follows that H_j is a major clique and contained in some grand clique K_j'. Since b_j and p are contained in K_j and K_j', they must coincide. Hence K_j contains H_j, which contradicts $|K_j| < |H_j|$.

Now consider the r grand cliques, K_1, K_2, ..., K_r. Then $K_1 - p$, $K_2 - p$, ..., $K_r - p$ are disjoint. For if $K_i - p$ and $K_j - p$, $i \neq j$, have a common vertex q, then K_i and K_j would coincide, and would contain both b_i and b_j, which is impossible since b_i is not adjacent to b_j. Remembering (5.4.3) we can now deduce

$$(5.4.3) \qquad \sum_{j=1}^{r} |K_j - p| \geqslant r(k-1-\alpha).$$

If possible, suppose there is another grand clique K_{r+1} containing p. The vertices in $K_{r+1} - p$ must be disjoint from the vertices in $K_1 - p$, $K_2 - p$, ..., $K_r - p$. Since K_{r+1} is a grand and therefore a major clique, $|K_{r+1} - p| \geqslant k - 1 - (r - 1)\alpha$. But from (c_1), the number of vertices adjacent to p is exactly $r(k - 1)$. Hence from (5.4.4)

$$r(k - 1) \geqslant \sum_{j=1}^{r+1} |K_j - p| > r(k - 1) + k - 1 - (2r - 1)\alpha.$$

$$\therefore k \leqslant 1 + (2r - 1)\alpha = q(r,\alpha) < \rho(r,\alpha,\beta),$$

which is a contrdiction. Thus p is contained in exactly r grand cliques.

5. **The fundamental characterization theorem.**

Theorem (5.5.1). Let G be a graph satisfying the conditions (c_1), (c_2), (c_3). Then if

$$k > \max[\gamma(r,\alpha), \rho(r,\alpha,\beta), p(r,\alpha,\beta)]$$

G is the graph of a WB–design with parameters (r, k, α, β).

If we take the vertices of G to be the treatments and the grand cliques of G to be the lines of the design then it follows directly from the theorems proved in the last two paragraphs that the design is a weakly balanced design with parameters (r, k, α, β).

The above is a slightly different version of the corresponding results in Bose and Lasker (1967), which may be consulted for further details of the proof.

R. C. BOSE

CHAPTER VI

SOME SPECIAL GRAPHS

1. **Pseudo geometric graphs.** A strongly regular graph is defined to be pseudo geometric (r, k, t) if its parameters v, n_1, p_{11}^1, p_{11}^2 are given by (4.4.7) and (4.4.4) where $1 \leq t \leq r$, $1 \leq t \leq k$. Thus a pseudo geometric graph (r, k, t) has the same parameters as the graph of a partial geometry (r, k, t) [Cf. paragraph 4, Chapter IV]. However, a pseudo geometric graph may not be the graph of a partial geometry. If G is the graph of a partial geometry (r, k, t), G may be called a geometric graph (r, k, t). We shall establish a sufficient condition for a pseudo geometric graph (r, k, t) to be geometric (r, k, t).

If G is a pseudo geometric graph (r, k, t), then it is a pseudo WB-graph with parameters (r, k, α, β), [Cf. Chapter V, paragraph 1], where

$$(6.1.1) \qquad \alpha = (r-1)(t-1), \quad \beta = rt$$

In this case the functions γ, q, ρ, p given by (5.2.1) - (5.2.4) become

$$(6.1.2) \qquad \gamma(r,t) = 1 + (r-1)^2(t-1) \ ,$$

$$(6.1.3) \qquad q(r,t) = 1 + (r-1)(2r-1)(t-1) \ ,$$

$$(6.1.4) \qquad \rho(r,t) = rt + (r-1)(2r-1)(t-1) \ ,$$

$$(6.1.5) \qquad p(r,t) = \tfrac{1}{2}[r(r-1) + t(r+1)(r^2 - 2r - 2)] \ .$$

Note that in view of the inequality $1 \leq t \leq r$ we have

$$(6.1.6) \qquad p(r,t) \geq \rho(r,t) \geq q(r,t) \geq \gamma(r,t) \ .$$

and

$$(6.1.7) \qquad p(r,t) > r$$

It follows from the fundamental characterization Theorem (5.5.1) that a pseudo geometric graph (r, k, t) is the graph of a WB—design D with parameters (r, k, α, β) if $k \geq p(r, t)$, where α and β are given by (6.1.1).

If we take the treatments of the design as points, the blocks of the designs as lines, and incidence as the relation of a treatment being contained in the design, then from the definition of a WB—design the axioms A1 and A2 for a partial geoemtry are satisfied. We shall show that axioms A3 and A4 are also satisfied.

Let θ be any treatment of D and let B_1, B_2, ..., B_r be the blocks containing θ. Since the average size of these blocks is k, the largest block B_j $(1 \leq j \leq r)$ contains at least k treatments. Let B be a subset of k treatments contained in B_j. Let $\bar{B}$ be the set of treatments not contained in B. Let g(x) be the number of treatments in $\bar{B}$ which have exactly x first associates in B. Then an easy counting argument shows that

$$\sum_{n=0}^{k} g(x) = k(k-1)(r-1)/t$$

and

$$\sum_{x=0}^{k} xg(x) = k(r-1)(k-1) .$$

Hence the average value of x is t and

$$\sum_{x=0}^{k} (x-t)^2 g(x) = 0$$

which is only possible if x is constant and equal to t. Hence every treatment in $\bar{B}$ has exactly t first associates in B.

If B_j contains any treatment φ other than those already contained in B, then φ belongs to $\bar{B}$ and therefore, has exactly t first associates in B.

R. C. BOSE

But each treatment of B is a first associate of ϕ. Hence
$t = k > p(r, t) > r$, which contradicts $t \leq r$.   This shows that each of the
block B_1, B_2, ..., B_r is of size k. Since θ is any arbitrary treatment
axiom A3 is satisfied. Also $x = t$ means that A4 is also satisfied.
Hence the design D is a partial geometry. We therefore have the following theorem [Bose (1963 c)].

Theorem (6.1.1). If the graph G is pseudo grometric (r, k, t) then
it is geometric if

$$k > p(r,t) = \frac{1}{2}[r(r-1) + t(r+1)(r^2 - 2r + 2)].$$

2.   Triangular and pseudo triangular graphs. The triangular association scheme with parameters given by (3.3.3) and (3.3.4) may be denoted
by $T_2(m)$. The corresponding strongly regular graph with paramenters

(6.2.1) $v = m(m-1)/2$, $n_1 = 2(m-2)$, $p_{11}^1 = m-2$, $p_{11}^2 = 4$,

will also be called a triangular or a $T_2(m)$ graph. The $m(m-1)/2$ treatments of a $T_2(m)$ association scheme may be made to correspond to the
$m(m-1)/2$ unordered pairs chosen out of a set of m symbols 1, 2, ..., m.
The pair (i, j) corresponds to the treatment occuring in the i-th row and
j-th column of the $m \times m$ square in which the treatments are exhibited.
Thus in the example m = 5, given in paragraph 3(b) of Chapter III, the
treatments 5, 8, 10 corresponds to the pairs (2,3), (3,4) and (4,5)
respectively. It is clear that if two treatments are first associates,
i.e., are in the same row or column, then the corresponding pairs have a
common symbol and if they are second associates, i.e., do not occur in
the same row or column, then the corresponding pairs have no symbol in
common. In the example under consideration, the treatments 5 and 8 corresponding to the pairs (2,3) and (3,4) are first associates, and the

treatments 5 and 10 corresponding to the pairs $(2,3)$ and $(4,5)$ are second associates. Hence we can label the vertices of a $T_2(m)$ graph G by
the $m(m-1)/2$ unordered pairs of symbols selected from 1, 2, ..., m.
Then two vertices are adjacent or non-adjacent according as the corresponding pairs have or do not have a common symbol.

A strongly regular graph G with the parameters (6.2.1) will be called
pseudo triangular or pseudo $T_2(m)$ graph. One may ask whether a pseudo
triangular graph is triangular, i.e., can the $m(m-1)$ vertices be labeled
by the unordered pairs (i,j), $1 \leq i$, $j \leq m$ such that two vertices are adjacent if and only if the corresponding pairs have a common symbol. The
answer is in the affirmative if $m \neq 8$. If $m = 8$ then besides the graph
$T_2(8)$ there are three other non-isomorphic graphs.

Now comparing (6.2.1) with (4.4.7) and (4.4.4) it is clear that a
pseudo triangular graph $T_2(m)$ is pseudo geometric $(2, m-1, 2)$. From
(6.1.5), $p(2,2) = 7$. Hence from Theorem (6.1.1) a pseudo $T_2(m)$ graph is
geometric $(2, m-1, 1)$ provided $m > 8$.

Thus a pseudo triangular graph $T_2(m)$, is the graph of a partial geometry $(2, m-1, 2)$. Let this graph be G. From the corollary to Theorem
(4.4.1), the number of lines in the geometry is m, and the number of
points is $m(m-1)/2$. Since $r = t = 2$, any two lines intersect in a unique
point. We may therefore label the lines by the symbols 1, 2, ..., m; and
the point of intersection of the line i and the line j, $i \neq j$ may be labeled by the unordered pair (i, j). It is now clear that two points are
incident with the same line if and only if the corresponding pairs have
a common symbol. This shows that G is a triangular graph.

We have thus shown that a pseudo triangular graph $T_2(m)$ is
triangular if $m > 8$. This result is due to Connor (1958). The same result

was proved by Shrikhande (1959 a) for $m \leq 6$, and by Hoffman (1960 a) for the case $m = 7$. However, it is surprising that when $m = 8$, the parameters (6.2.1) do not characterize a $T_2(8)$ graph. It turns out that there are three other non-isomorphic graphs. This was demonstrated by Hoffman (1960 b) and by Chang [(1959), (1960)].

Hoffman's proofs use the fact that if G is a strongly regular graph with parameters given by (6.2.1), then the characteristic roots of its adjacency matrix are

(a) $2m - 4$ with multiplicity 1, and characteristic vector (1, 1, ..., 1).

(b) $m - 4$ with multiplicity $m - 1$.

(c) -2 with multiplicity $v - m$.

Shrikhande's and Chang's proofs use purely combinatorial arguments. We can summarize these results in the following theorem.

Theorem (6.2.1). If G is a pseudo triangular graph with parameters given by (6.2.1), then it is triangular if $m \neq 8$. When $m = 8$ then there are three other non-isomorphic graphs besides $T_2(8)$.

It is of interest to examine the nature of the other three pseudo triangular graphs $T_2(8)$ with parameters

$$v = 28, \quad n_1 = 12, \quad p_{11}^1 = 6, \quad p_{11}^2 = 4 .$$

The condition (b) of Theorem (4.2.2) is satisfied since

$$p_{11}^1 + p_{11}^2 = 2n_1 - \frac{v}{2} = 10$$

We start with the graph $T_2(8)$, whose vertices are the unordered pairs which can be formed from the symbols 1, 2, ..., 8 and in which two vertices are adjacent if and only if the corresponding pairs have a common symbol. Theorem (4.2.2) now shows that if we obtain $T_2*(8)$ from $T_2(8)$ by

R. C. BOSE

complimentation with respect to V_1, where V_1 is any of the three sets, S_1, S_2 or S_3 and $V_2 = V - V_1$; and V is the set of vertices of $T_2(8)$, then $T_2^*(8)$ has the same parameters as $T_2(8)$. The sets S_1, S_2, S_3 are given below

$S_1 = \{(12),(34),(56),(78)\}$,

$S_2 = \{(12),(34),(56),(78),(13),(24),(57),(68)\}$,

$S_3 = \{(12),(34),(56),(78),(13),(24),(57),(68),(14),(23),(58),(67)\}$.

$T_2(8)$ cannot have any claw of order 3, and there is no complete quadrangle. The non-isomorphism of the other three graphs can be checked by counting claws of order 3, and the number of complete quadrangles.

3. **The Hall-Connor embedding theorem.** Consider a symmetric BIB design with parameters $v = b$, $r = k$, λ. We have shown in Theorem (1.5.1) that we can obtain from it another BIB design D^*, the residual of D by deleting one block and all the treatments contained in this block. The parameters of D^* are

(6.3.1) $v^* = v - k$, $b^* = b - 1$, $k^* = k - \lambda$, $r^* = r$, $\lambda^* = \lambda$.

One may ask the following question: Given a BIB design D^* with parameters (6.3.1), is it possible to obtain from it a symmetric BIB design with parameters $v = b$, $r = k$, λ by taking an additional block of k new treatments and adjoing to each of the blocks of D^*, λ suitably chosen treatments out of the k new treatments?

When $\lambda = 1$, D^* is isomorphic to an affine plane, and can be converted to a projective plane (isomorphic to D) in the usual manner. Hence for $\lambda = 1$ the result is certainly true. Hall and Connor (1953) showed that the result is also true for $\lambda = 2$. For the case $\lambda = 3$ a counter example was given by Bhaltacharye (1944).

Shrikhande (1960) gave an alternative proof of the Hall-Connor theorem depending on the characterization theorem (6.2.1) for pseudo triangular graphs.

For the case $\lambda = 2$, the parameters of the residual design D^* can be written as

(6.3.1) $v^* = \dfrac{(k-1)(k-2)}{2}$, $b^* = \dfrac{k(k-1)}{2}$, $r^* = k$, $\lambda^* = k-2$, $\lambda^* = 2$.

It is readily proved that two blocks of D intersect in either one or two treatments. If we take the blocks of D^* as the treatments of an association scheme and call two blocks first associates or second associates according as they intersect in one treatment or two treatments then it can be shown [Hall and Connor (1953), Shrikhande (1952)] that we get a two class association scheme with parameters

(6.3.2) $v = k(k-1)/2$, $n_1 = 2(k-2)$, $p_{11}^1 = k-1$, $p_{11}^2 = 4$.

Hence the corresponding strongly regular graph is the pseudo triangular graph $T_2(k)$, which by Theorem (6.2.1) is triangular $T_2(k)$ if $k \neq 8$. If we now take k symbols 1, 2, ..., k then to each block of D^* we can assign an unordered pair (i j) of symbols such that two blocks are first associates, i.e. intersect in one treatment if and only if the corresponding pairs have one symbol in common. We can now take k new treatments. If (i j) is the pair assigned to the block B_{ij} we extend it by adding two new treatments i and j. Finally, we add a new block consisting of all the new treatments. We have now obtained a design with $(k^2 - k + 2)/2$ blocks and treatments, with each treatment appearing in k blocks of size k, such that any two blocks intersect in two treatments. From this it follows that every pair occurs in two blocks. This proves

Hall and Connors theorem if $k \neq 8$. Connor (1952) gave a separate proof
for the non-existence of (6.3.1) for the case $k = 8$. Hence the problem
of embedding does not arise.

4. <u>Embedding the complement of an oval in a projective plane of even</u>
<u>order</u>. Given a projective plane of even order q, it is well known [Bose
(1947), Segre (1954) and (1955)] that we can find a set of $q + 2$ points in
Π such that no three are collinear. They form an oval in Π. If we delete
the points of the oval from Π, then the incidence structure of the remaining points and lines has the following properties:

(a) There are $q^2 - 1$ points.

(b) There are two types of lines. Lines of type I are each incident
with $q + 1$ points. Lines of type II are each incident with $q - 1$ points.

(c) Each point is incident with $q/2$ lines of type I and $(q + 2)/2$
points of type II.

(d) Any two distinct points are both incident with exactly one line
which may be of type I or type II.

Conversely let D be an incidence structure with the above properties.
We may ask the question whether it is possible to embed D in a projective
plane of order q by suitably extending the lines of type II. Bose and
Shrikande (1972) have proved that the answer is in the affirmative except
possibly for the case $q = 6$. Their proof essentially depends on showing
that if we take the lines of type II as the treatments of an association
scheme and call the two lines first associates if they do not intersect
and second associates if they intersect in a point, then the association
scheme has the parameters (6.2.1) with $m = q + 2$. Hence from Theorem
(6.2.1) the corresponding graph is triangular except possibly in the case
$q = 6$. This enables them to suitably extend the lines.

5. **Net and pseudo net graphs.** A net (r, k) of degree r and order k is a system of undefined points and lines together with an incidence relation subject to the following axioms (i) There is at least one point (ii) The lines of the net can be partitioned into r disjoint, nonempty, "parallel classes" such that each point of the net is incident with exactly one line of each class and given two lines belonging to distinct classes there is exactly one point of the net which is incident with both lines (iii) One line is incident with k points.

For convenience we can use phrases such as "point is on a line" instead of speaking incidence. Then it can be readily proved [Bruck (1963)] that

(1) Each line of the net contains exactly k distinct points where $k \geq 1$.

(2) Each point of the net lies on exactly r distinct lines where $r \geq 1$.

(3) The net has exactly rk distinct lines. These lines fall into r parallel classes of k lines each. Distinct lines of the same parallel class have no common points. Two lines of different classes have one common point.

(4) The net has exactly k^2 distinct points.

It is easy to show [Bose (1963 c)] the equivalence of a net (r, k), a partial geometry $(r, k, r-1)$ and a set of $r-2$ mutually orthogonal Latin squares of order k.

The graph of a net (r, k) or the corresponding partial geometry $(r, k, r-1)$ can be defined as usual. The points of the net or the partial geometry correspond to the vertices of the graph and two vertices are adjacent if and only if the corresponding points are incident with the same

R. C. BOSE

For $r \geq 3$ very little is known as to what happens when $k \leq p(r,r-1)$. It is an open question whether $p(r,r-1)$ is the best possible value for the result to hold, i.e. does there always exist a graph non-isomorphic to the net graph $L_r(k)$ when $k = p(r,r-1)$.

6. **The Bruck-Shrikhande embedding theorem.** Bruck defines the deficiency d of a net (r, k) by

$$(6.6.1) \qquad\qquad d = k - r + 1 .$$

The interpretation of the deficiency d is that if it were possible to add d more parallel classes, each consisting of k lines, so that the extended net now has $k+1$ classes of parallels, the net would become an affine plane, in which any two points are joined by a unique line. We shall now prove:

Theorem (6.6.1). A net (r, k) of deficiency d can be completed to an affine plane of order k if

$$(6.6.2) \qquad\qquad k > \frac{1}{2}(d-1)(d^3 - d^2 + d + 2) .$$

From the equivalence of a set of $r-2$ mutually orthogonal Latin squares of order k and a net (r, k) we can write the above theorem in the following equivalent form

Theorem (6.6.2). If there exist $k-1-d$ mutually orthogonal Latin squares of order k, it is possible to get a complete set of $k-1$ mutually orthogonal Latin squares, by adding d new suitably chosen squares, provided that $k > \frac{1}{2}(d-1)(d^3 - d^2 + d + 2)$.

The case $d = 2$, was first obtained by Shrikhande (1961) and the general case was obtained by Bruck (1963). When $d = 2$, the theorem holds for $k > 4$, but $k = 4$ is exceptional as it is well known that a cyclic Latin square of order 4 cannot belong to a complete set of three mutually

line. It follows from Theorem (4.4.1) and can easily be verified directly that the parameters of the graph of a net (r, k) are

(6.5.1) $v = k^2$, $n_1 = r(k-1)$, $p^1_{11} = (r-2)(r-1)+k-2$, $p^2_{11} = r(r-1)$

The graph of a net (r, k) will be called a net graph $L_r(k)$. A graph
with paramenters (6.5.1) will be called a pseudo net graph $L_r(k)$. A
pseudo net graph $L_r(k)$ is not necessarily a net graph. However, as a
consequence of Theorem (6.1.1), we have

Theorem (6.5.1). A pseudo net graph $L_r(k)$ is a net graph $L_r(k)$ if

$$k > p(r,r-1) = \frac{1}{2}(r-1)(r^3-r^2+r+2) .$$

The special case r = 2 was first proved by Shrikhande (1959 b). In
this case if k > 4, then a pseudo net graph $L_2(k)$ is a net graph $L_2(k)$.
It is easy to check that this result is also true for k < 4. But the case
k = 4 is exceptional.

For a net (2, k) a point P may be given coordinates (i, j),
$1 \leqslant i,j \leqslant k$ if it is on the i-th line of the first parallel class and j-th
line of the second parallel class. Then vertices of the net graph $L_2(k)$
are the ordered pairs (i, j) two vertices being adjacent if and only if
they have a common coordinate. The condition (b) of Theorem (4.2.2) is
satisfied if k = 4. The condition (a) is also satisfied if we take

V_1=(1,1),(2,2),(3,3),(4,4)

V_2=(1,2),(1,3),(1,4),(2,1),(2,3),(2,4),(3,1),(3,2),(3,4),(4,1),(4,2),(4,3).

The graph G* obtained by complementation with respect to V_1 and V_2 has
the parameters (6.5.1) for k = 2, but is non-isomorphic to the net graph
$L_2(4)$.

The general case of Theorem (6.5.1) was obtained by Bruck (1963).

orthogonal Latin squares.

The proof of these two theorems depends on Theorem (6.5.1). Let G be a pseudo net graph $L_r(k)$. Then its parameters are given by (6.5.1). The other parameters n_2 and p^i_{jk} (i, j, k = 1, 2) are given by (3.3.12), (3.3.13) and (3.3.14). Let $\bar{G}$ be the complementary of G, i.e. the vertices of $\bar{G}$ are the same as the vertices of G but the vertices adjacent in G are non-adjacent in $\bar{G}$ and vise-versa. Then the parameters of $\bar{G}$ are obtained from G by reversing the subscripts 1 and 2. Thus

$$\bar{n}_1 = n_2 = (k-1)(k-r+1) = d(k-1)$$

$$\bar{p}^1_{11} = p^2_{22} = (k-r)^2 + (r-2) = (d-2)(d-1) + (k-2)$$

$$\bar{p}^2_{11} = p^1_{22} = (k-r)(k-r+1) = d(d-1) .$$

Hence $\bar{G}$ is strongly regular with parameters

$$\bar{v} = k^2, \quad \bar{n}_1 = d(k-1), \quad \bar{p}^1_{11} = (d-1)(d-2) + (k-2), \quad \bar{p}^2_{11} = d(d-1)$$

Thus $\bar{G}$ is pseudo $L_d(k)$. From Theorem (6.5.1) it is the net graph $L_d(k)$ if (6.6.2) is satisfied. Thus there exists a net (d, k) with the same points as the net (r, k) but if two points are incident or non-incident with the same line in (r, k), they they are non-incident or incident with the same line in (d, k). Adding the d k lines of the net (d, k) to the (r, k) lines of the net (r, k) we have a net (k+1, k) which is an affine plane of order k. This proves the required result.

7. $\underline{T_q(m)}$ and pseudo $\underline{T_q(m)}$ graphs. Consider a set S of m symbols 1, 2, ..., m where m ≥ 2. We can form $\binom{m}{q}$ unordered q-plets of q distinct elements from S, 1 ≤ q ≤ m. Let G be a graph whose vertices are the unordered q-plets of S. Let two vertices of G be adjacent if the corresponding q-plets have exactly q - 1 symbols of S in common, and non-adjacent

otherwise. We shall call G a $T_q(m)$ graph. Clearly G has the following properties:

(1) G is regular of valence $d = q(m-q)$.

(2) G is edge regular with edge degree $\delta = m-2$.

(3) If x and y are non-adjacent vertices of G then $\Delta(x, y) \leq 4$.

A graph with $\binom{m}{q}$ vertices, and satisfying the conditions (1), (2) and (3) will be called pseudo $T_q(m)$ graph. We can then ask under what conditions a pseudo $T_q(m)$ graph is a $T_q(m)$ graph.

Now the conditions (1), (2) and (3) are the same as the conditions (c_1), (c_2) and (c_3) of Chapter V, paragraph 1, provided we set

(6.7.1) $\qquad r = q, \quad k = m-q+1, \quad \alpha = q-1, \quad \beta = 4$.

Also d, δ, r are positive integers, β is a non-negative integer and $k > 2$, $\alpha > 0$. Hence a pseudo $T_q(m)$ graph is a pseudo WB-graph.

The functions $\gamma(r,\alpha)$, $q(r,\alpha)$, $\rho(r,\alpha,\beta)$, $p(r,\alpha,\beta)$ given by (5.2.1) - (5,2.4) now become

(6.7.2) $\qquad \gamma(r,\alpha) = 1 + (q-1)^2$,

(6.7.3) $\qquad q(r,\alpha) = 1 + (2q-1)(q-1)$,

(6.7.4) $\qquad \rho(r,\alpha,\beta) = 4 + (2q-1)(q-1)$,

(6.7.5) $\qquad p(r,\alpha,\beta) = 1 + \frac{1}{2}(q+1)(q+2)$.

Clearly $\max[\gamma(r,\alpha), \rho(r,\alpha,\beta), p(r,\alpha,\beta)] = 4 + (2q-1)(q-1)$.

Hence from the fundamental characterization theorem a pseudo $T_q(m)$ graph G is the graph of a WB-design with parameters (r, k, α, β) given by (6.7.1), if $k > 4 + (2q-1)(q-1)$, i.e.

(6.7.6) $\qquad m > 4 + 2q(q-1)$.

The treatments of the WB-design correspond to the vertices of G and

R. C. BOSE

the blocks of the WB–design corresponds to the grand cliques of G. Bose and Lasker (1967) showed for the case m = 3 and Dowling (1969) for the general case that if we take the grand cliques of G as the vertices of a new graph G^* and consider two vertices of G^* adjacent if the corresponding grand cliques both contain a vertex of G in common then G^* satisfies the conditions (1), (2) and (3) if we replace m by $m^* = m - 1$. Using induction we have

Theorem (6.7.1). If for a graph G with $\binom{m}{q}$ vertices the conditions (1), (2) and (3) are satisfied, i.e. G is a pseudo $T_q(m)$, then G is a $T_q(m)$ graph if

$$m > 4 + 2q(q - 1)$$

Notice that the case q = 2 reduces to Connor's result that for m > 8 a pseudo triangular graph $T_2(m)$ is triangular $T_2(m)$.

In particular if a graph G with $m(m-1)(m-2)/6$ vertices satisfies the conditions (1), (2) and (3) with q = 3 then G is a $T_3(m)$ graph if m > 16. Aigner (1969) has shown that the same hold if m ⩽ 9. The question is open for 9 ⩽ m ⩽ 16 though no exceptional cases are known. For q > 3 nothing is known about the case when m ⩽ 2q(q - 1) + 4.

8. Characterization of cubic lattice graphs. A cubic lattice graph of order m is a graph G whose vertices can be identified with the ordered triplets on m symbols so that two vertices are adjacent if the corresponding triplets have common symbols in exactly two positions. If G is a cubic lattice graph of order m then G has m^3 vertices and possesses the following properties.

(i) G is regular with valence d = 3(m - 1).

(ii) G is edge regular with edge degree $\delta = m - 2$.

R. C. BOSE

(iii) If x and y are non-adjacent vertices of G, then $\Delta(x, y) = 2$, if $d(x, y) = 2$, and $\Delta(x, y) = 0$ if $d(x, y) > 2$, where $d(x, y)$ denotes the distance between x and y.

Lasker (1967) and Dowling (1968) have proved

Theorem (6.8.1). A graph G with m^3 vertices and possessing the properties (i), (ii) and (iii) is a cubic lattice graph of order m if $m > 7$.

Lasker's original characterization had an additional assumption which was eliminated by Dowling. The starting point of their proofs is that the properties (i), (ii) and (iii) imply the properties (c_1), (c_2) and (c_3) of Chapter V, paragraph 1 if

(6.8.1) $\qquad\qquad r = 3, \quad k = m, \quad \alpha = 0, \quad \beta = 2 .$

Hence

$$\gamma(r,\alpha) = 1, \quad q(r,\alpha) = 1, \quad \rho(r,\alpha,\beta) = 2, \quad p(r,\alpha,\beta) = 7$$

Therefore the fundamental characterization theorem (5.5.1) applies if $m > 7$.

CHAPTER VII

GRAPHS IN WHICH EACH PAIR OF VERTICES IS ADJACENT TO
THE SAME NUMBER d OF OTHER VERTICES

1. Graphs for which p_{11}^1 and p_{11}^2 are constant. Let G be a finite graph not necessarily connected. If x and y are any two vertices let $\Delta(x, y)$ denote the number of vertices simultaneously adjacent to both x

R. C. BOSE

and y. Bose and Dowling (1971) proved

Theorem (7.1.1). If for a finite graph G, $\Delta(x, y) = p_{11}^1$ if x and y are adjacent, and $\Delta(x, y) = p_{11}^2$ if x and y are not adjacent, where p_{11}^1 and p_{11}^2 are fixed non-negative integers, then only the following cases are possible

    (i) G is regular and hence strongly regular.

    (ii) $p_{11}^2 = 1$, and G consists of n complete subgraphs of order $p_{11}^1 + 2$, with one common vertex.

    (iii) $p_{11}^2 = 0$, and G consists of n disjoint complete subgraphs of order $p_{11}^1 + 2$, and m isolated points.

The proof depends on simple counting arguments and a number of Lemmas on the existence of 3 and 4 cycles in the graphs.

    2. $G_2(d)$ graphs and symmetric BIB designs. A finite graph G is defined to be a $G_2(d)$ graph if any two distinct vertices x, y are both adjacent to exactly d other vertices. Hence $\Delta(x, y) = d$ whether x and y are adjacent or non-adjacent. Thus Theorem (7.1.1) applies with $p_{11}^1 = p_{11}^2 = d$. Thus a $G_2(d)$ graph is necessarily regular if $d \geqslant 2$. Again from the same theorem we have

    Lemma (7.2.1). An irregular $G_2(d)$ graph with $d = 1$, must consist of $n \geqslant 2$ complete graphs of order 3 with a common vertex.

The case $d = 0$ is trivial. It is clear that in this case the graph consists of m isolated vertices, and n disjoint edges together with their 2n vertices.

We can therefore confine our attention to regular $G_2(d)$ graphs. A regular $G_2(d)$ graph with v vertices and valence n_1 will be said to have parameters (v, n_1, d). We can obtain from this a symmetric BIB design for which $r = k = n_1$, $\lambda = d$ in the following manner: Let the vertices of the

R. C. BOSE

graph be taken as treatments. Let the i-th block consist of all treatments which correspond to vertices adjacent to the i-th vertex. Then
clearly each block is of size n_1 and each treatment occurs in n_1 blocks.
Also any two treatments x and y will both occur in the block i if and
only if the vertex i is adjacent to both the vertices x and y. Hence
any two treatments occur together in d blocks. It follows from (1.1.1)
that $v = b = n_1(n_1 - 1)/d$. This BIB design has the property (S) that its
incidence matrix $N = (n_{ij})$ can be written in a form such that $n_{ij} = n_{ji}$
for $i \neq j$ and $n_{ii} = 0$. A symmetric BIB design with v treatments, block
size k and for which every pair of treatments occur in λ block will be
called a symmetric (v, k, λ) BIB design. It is clear that from a symmetric (v, k, λ) BIB design, with the property (S), we can obtained a $G_2(\lambda)$
graph with parameters (v, k, λ). Hence we have

Theorem (7.2.1). The existence of a $G_2(d)$ graph with parameters
(v, n_1, d) is equivalent to the existence of a symmetric (v, n_1, d) BIB
design with the property (S).

Corollary. The parameters (v, n_1, d) of a regular $G_2(d)$ graph are
connected by the relation

(7.2.1) $$v - 1 = n_1(n_1 - 1)/d .$$

3. Necessary conditions for the existence of regular $G_2(d)$ graphs.
Let A be the adjacency matrix of a regular $g_2(d)$ graph with parameters
(v, n_1, d). Let J be the $v \times v$ matrix for which each element is unity.
Then it is readily seen that

$$A^2 = (n_1 - d)I + dJ .$$

Hence

$$A^3 = (n_1 - d)A + n_1 dJ .$$

$$A^3 - n_1 A^2 - (n_1 - d)A + n_1(n_1 - d)I = 0 .$$

Thus A has only three distinct characteristic roots n_1 and $\pm (n_1 - d)^{1/2}$. Now from the regularity of $G_2(d)$ it follows that $A^* = A/n_1$ is a stochastic matrix, and since $G_2(d)$ is connected A^* is irreducible. It follows [Brauer (1952)] that unity is a simple root of A^*, so that n_1 is a simple root of A. Let α_1, α_2 be the multiplicities of the roots $\theta_1 = (n_1 - d)^{1/2}$ and $\theta_2 = -(n_1 - d)^{1/2}$. Then

$$|A - I\theta| = (n_1 - \theta)(\theta_1 - \theta)^{\alpha_1}(\theta_2 - \theta)^{\alpha_2} .$$

To determine α_1 and α_2 we note that

$$\text{Tr } I = 1 + \alpha_1 + \alpha_2 = v ,$$

$$\text{Tr } A = n_1 + \alpha_1\theta_1 + \alpha_2\theta_2 = 0 .$$

$$\therefore \ \alpha_1 = \frac{v-1}{2} + \frac{n_1}{2(n_1 - d)^{1/2}}, \quad \alpha_2 = \frac{v-1}{2} - \frac{n_1}{2(n_1 - d)^{1/2}} .$$

Since the multiplicities are necessarily integral, there must exist an integer m such that $n_1 = d + m^2$. Also since

$$2\alpha_2 = (v - 1) - (m + \frac{d}{m})$$

d/m must be integral, and the integers $v - 1 - m$ and d/m must have the same parity.

In particular let d = 1. Then m = 1 and from (7.2.1), v = 3. Hence a regular $G_2(d)$ graph with d = 1 must necessarily be a complete graph of order 3. Taking this together with Lemma (7.2.1) we have the theorem of Erdős, Renyi and Sós (1966).

Theorem (7.3.1). A finite graph G in which any two distinct vertices are simultaneously adjacent to exactly one vertex, consists of n sub-

R. C. BOSE

graphs of order 3, which have a common vertex when $n \geqslant 2$.

Thus $G_2(d)$ graphs with $d = 0$ or 1 are completely characterized. For $d \geqslant 2$ we have the following theorem:

Theorem (7.3.2). If $G_2(d)$ is a finite graph without loops or multiple edges, in which each pair of distinct vertices is adjacent to exactly d other vertices, $d \geq 2$, then $G_2(d)$ is regular of valence n_1 such that $v - 1 = n_1(n_1 - 1)/d$ where v is the number of vertices and there exists a positive integer m, such that

(i) $n_1 = d + m^2$

(ii) d/m is an integer, with the same parity as $v - 1 - m$.

We will next address ourselves to the problem of construction of $G_2(d)$ graphs.

4.   <u>$G_2(d)$ graphs derived from partial geometries.</u> Consider a pseudo geometric graph (r, k, t), with parameters v, n_1, p_{11}^1, p_{11}^2 given by (4.4.7) and (4.4.4). When the condition $k = r + t + 1$ is satisfied we have $p_{11}^1 = p_{11}^2 = rt$. Since the graph of a partial geometry is also pseudo geometric any partial geometry for which $k = r + t + 1$ will provide a $G_2(d)$ graph. Many examples of partial geometries satisfying these conditions are known.

(a)   Consider an elliptic non-degenerate quadric in the finite projective space PG(5, q) where q is a prime power. It is known [Primrose (1957) and Ray-Chaudhuri (1961 a)] that this quadric is ruled by straight lines called generators, but contains no planes. The number of points on the surface is $(q^3 + 1)(q^2 + 1)$. It was shown in Bose (1963 c), that the points and generators can be regarded as the points and lines of a partial geometry $(q^2 + 1, q + 1, 1)$. The dual of this partial geometry is obtained by taking the points and lines of the dual to be the line and points of the original geometry. Thus the dual partial geometry has the parameters $(q+1, q^2+1, 1)$. Another way to obtain a partial geometry with the same parameters is to take the surface $x_0^{q+1} + x_1^{q+1} + x_2^{q+1} + x_3^{q+1}$ = 0 in $PG(3, q^2)$. It has been shown in Bose and Chakravarti (1966) that this surface contains $(q^2+1)(q^3+1)$ points, and $(q+1)(q^3+1)$ generators which may be taken to be points and lines of a partial geometry $(q+1, q^2+1, 1)$. If we take $q = 2$, the condition $p_{11}^1 = p_{11}^2$ is satisfied. The graph of this partial geometry is strongly regular with parameters $(45, 12, 3, 3)$. Thus we get a $G_2(d)$ with $v = 45$, $n_1 = 12$, $d = 3$. Here $m = 3$.

(b) For a pseudo net graph $L_r(k)$ with parameters (6.5.1) the condition $k = r + t + 1$ reduces to $k = 2r$. Thus a pseudo net graph $L_r(2r)$ is a $G_2(d)$ graph with $v = 4r^2$, $n_1 = r(2r-1)$, $d = r(r-1)$. We can thus get $G_2(d)$ graphs for all values of r for which there exist $r-2$ mutually orthogonal Latin squares of order $2r$. This is always true if $r = 2^m$ where m is a non-negative integer. We can therefore obtain a $G_2(d)$ graph with $v = 2^{2m+2}$, $n_1 = 2^m(2^{m+1}-1)$, $d = 2^m(2^m-1)$. Again since there exists a Latin square of order 6, we get a $G_2(d)$ graph with $v = 36$, $n_1 = 15$, $d = 6$. Also the existence of 5 mutually orthogonal squares of order 12 is known [Bose, Chakravarti, and Knuth (1960)]. By taking 4 mutually orthogonal squares of order 12, we can get a $G_2(d)$ graph with $v = 144$, $n_1 = 66$, $d = 30$.

(c) The dual of a design is defined as a new design whose treatments and blocks are in (1, 1) correspondence with the blocks and treatments of the original design, and incidence is preserved (a block and treatment are incident if the treatment is contained in the block and non-incident otherwise). It is known [Bose (1963 c)] that the dual of a BIB design with parameters v_0, b_0, r_0, k_0, $\lambda_0 = 1$ can be regarded as partial geometry (r, k, r) where $r = k_0$, and $k = r_0$, the lines of the partial geometry being the blocks of the dual design.

Such a dual design (or partial geometry) may be called a linked block design (or geometry). The corresponding strongly regular graph has the parameters

(7.4.1) $$v = k(kr - k + 1)/r, \quad n_1 = r(k - 1) ,$$

(7.4.2) $$p_{11}^1 = (k - 2) + (r - 1)^2, \quad p_{11}^2 = r^2 .$$

It will be called a linked block graph and will be denoted by $LB_r(k)$. Any strongly regular graph (not necessarily the graph of a linked block design) will be called a pseudo linked block graph $LB_r(k)$. The condition $k = r + t + 1$ now reduces to $k = 2r + 1$. Thus a pseudo linked block graph $LB_r(2r + 1)$ is a $G_2(d)$ graph with $v = 4r^2 - 1$, $n_1 = 2r^2$, $d = r^2$. Now BIB designs with parameters $v_0 = 2^{m-1}(2m - 1)$, $b_0 = 2^{2m} - 1$, $r_0 = 2^m + 1$, $k_0 = 2^{m-1}$, $\lambda_0 = 1$ are known [Bose and Shrikhande (1960 b)] for every integral value of m. We can therefore get a corresponding $G_2(d)$ graph with $v = 2^{2m} - 1$, $n_1 = 2^{2m-1}$, $d = 2^{2m-2}$ for all integral m. Also BIB designs with parameters

(7.4.3) $v_0 = r(2r - 1)$, $b_0 = 4r^2 - 1$, $r_0 = 2r + 1$, $k_0 = r$, $\lambda_0 = 1$

are known for values of $k_0 = 2, 3, 4, 5$ and 7 [Bose (1939) and Rao (1961)]. Hence the corresponding $G_2(d)$ graphs with parameters $v = 4r^2 - 1$, $n_1 = 2r^2$, $d = r^2$ can be constructed for $r = 2, 3, 4, 5$ and 7.

5. <u>$G_2(d)$ graphs of negative Latin square type.</u> A strongly regular graph corresponding to Mesner's negative Latin square association scheme [Chapter III, paragraph 3(e)] will be called a negative Latin square

graph $NL_r(k)$. Its parameters are

(7.5.1) $$v = k^2, \quad n_1 = r(k+1) ,$$

(7.5.2) $$p^1_{11} = (r+1)(r+2) - (k+2), \quad p^2_{11} = r(r+1) .$$

If $k = 2r$, then $p^1_{11} = p^2_{11} = r(r-1)$. Hence a negative Latin square graph $NL_r(2r)$ is a $G_2(d)$ graph with parameters $v = 4r^2$, $n_1 = r(2r+1)$, $d = r(r+1)$.

We shall now show that a negative Latin square graph $NL_r(2r)$ can be derived from a net graph $L_r(2r)$. A net graph $L_r(2r)$ is the graph of a net of degree r and order $2r$. Take any class C of parallel lines in the net, and divide them into groups of r lines each. Let V_1 be the set of vertices corresponding to the $2r^2$ points on lines of the first group, and V_2 be the set of vertices corresponding to the $2r^2$ points on the lines of the second group. If P is a point on a line ℓ of the first group, then the vertex x corresponding to P is adjacent to the $2r-1$ vertices corresponding to the other points on ℓ. Also through P there pass $r-1$ lines other than ℓ (one belonging to each of the parallel classes other than C). Each of these lines intersects each line of the first group (other than ℓ) in a single point. The vertices corresponding to these $(r-1)^2$ intersections are adjacent to the vertex x. We thus get $w_1 = r^2$ vertices in V_1 adjacent to x. It is clear that these are all the vertices in V_1 adjacent to x. Similarly each vertex in V_2 is adjacent to exactly $w_2 = r^2$ vertices in V_2. Again for the net graph $L_r(2r)$, $2n_1 - (v/2) = p^1_{11} + p^2_{11} = 2r(r-1)$. Thus the conditions of Theorem (4.2.1) are satisfied. Hence by complementation with respect to V_1 and V_2 we obtain a strongly regular graph with parameters

$$v^* = 4r^2, \quad n^*_1 = r(2r+1) ,$$

$$p_{11}^{1*} = r(r+1) = p_{11}^{2*} .$$

This is a negative Latin square graph $NL_r(2r)$ by definition. We thus have

Theorem (7.5.1). A negative Latin square graph $NL_r(2r)$ exists whenever a net graph $L_r(2r)$ exists. In particular a negative Latin square graph $NL_r(2r)$ exists for $r = 3$, 6 and 2^{n-1} where $n > 1$.

6. Composition of pseudo net and pseudo negative Latin square graphs. We have seen that when the incidence matrix $N = (n_{ij})$ of a symmetric BIB design (v_0, k_0, λ_0) satisfies the conditions (S), $n_{ij} = n_{ji}$ for $i \neq j$, $n_{ii} = 0$ then N is the adjacency matrix of a $G_2(\lambda_0)$ graph. We will call N, an incidence matrix of type I.

Let J denote the $v_0 \times v_0$ matrix all of whose elements are unity, If N is the incidence matrix of a symmetric BIB design (v_0, k_0, λ_0), and $\overline{N} = \overline{J} - N$ then $\overline{N}$ is the incidence matrix of a symmetric BIB design $(v_0, v_0 - k_0, v_0 - 2k_0 + \lambda_0)$. If N is of type I, then $\overline{N}$ is symmetric, and each element of the main diagonal is unity. The incidence matrix of a symmetric BIB will be defined to be of type II when it satisfies these conditions.

Consider a class Ω of symmetric BIB designs (v_0, k_0, λ_0) for which the condition $v_0/4 = k_0 - \lambda_0$ is satisfied. Then the following result which we state in the form of a lemma is known [Shrikhande (1962)].

Lemma (7.6.1). If N_1 and N_2 are the incidence matrices of two symmetric BIB designs (v_1, k_1, λ_1) and (v_2, k_2, λ_2) belonging to the class Ω, then

$$N = N_1 \times N_2 + \overline{N}_1 \times \overline{N}_2$$

is the incidence matrix of a symmetric BIB design (v_0, k_0, λ_0) belonging

R. C. BOSE

to Ω where

$$v_0 = v_1 v_2, \quad k_0 = k_1 k_2 + (v_1 - k_1)(v_2 - k_2), \quad \lambda_0 = k_0 - (v_0/4) ,$$

and $\times$ denotes the Kronecker product.

We also note that if N_1 is of type II and N_2 is of type I, then N is of type I.

If the adjacency matrix of a $G_2(d)$ graph with parameters v, n_1, d is the incidence matrix of a BIB design of the class Ω, then $v = r(n_1 - d)$. Since $v - 1 = n_1(n_1 - 1)/d$, it follows that $n_1 = \frac{1}{2}[(4d + 1) \pm (4d + 1)^{1/2}]$, which shows that $n_1 = r(2r + 1)$ or $n_1 = r(2r - 1)$ for some positive integer r. Thus $G_2(d)$ must be either a pseudo net graph $L_r(2r)$ or a negative Latin square graph $L_r(2r)$.

Theorem (7.6.1). The existence of pseudo net graphs $L_{r_1}(2r_1)$ and $L_{r_2}(2r_2)$ implies the existence of a pseudo net graph $L_r(2r)$ with $r = 2r_1 r_2$.

Let N_1 be the adjacency matrix of the pseudo net graph $L_{r_1}(2r_1)$ and N_2 the adjacency matrix of the pseudo net graph $L_{r_2}(2r_2)$. Then N_1 is the incidence matrix of a symmetric BIB $[4r_1^2, r_1(2r_1 - 1), r_1(r_1 - 1)]$ and $\overline{N}_2$ is the adjacency matrix of a symmetric BIB $[4r_2^2, r_2(2r_2 + 1), r_2(r_2 + 1)]$, where N_1 is to type I and $\overline{N}_2$ is of type II. From Lemma (7.6.1)

$$N = \overline{N}_2 \times N_1 + N_2 \times \overline{N}_1$$

is the adjacency matrix of a symmetric BIB (v_0, k_0, λ_0), belonging to Ω, where

$$v_0 = 16r_1^2 r_2^2, \quad k_0 = 2r_1 r_2(4r_1 r_2 - 1) ,$$

$$\lambda_0 = (2r_1 r_2 - 1) .$$

Since N is of type I, it follows that it is the adjacency matrix of a pseudo net graph $L_r(2r)$, where $r = 2r_1 r_2$.

Theorem (7.6.2). A pseudo net graph $L_r(2r)$ exists for all $r = 3^m \cdot 2^{m+n-1}$, where m and n are non-negative integers, $(m, n) \neq (0, 0)$.

A net graph is also a pseudo net graph. Hence a pseudo net graph $L_r(2r)$ exists for $r = 2^{n-1}$, $n \geqslant 1$. Hence the theorem is true for $m = 0$, $n \geqslant 1$. Again a net graph $L_r(2r)$ exists for $r = 3 \cdot 2^0$. Assuming the existence of a pseudo net graph $L_r(2r)$ for $r = 3^{m-1} \cdot 2^{m-2}$, the existence of a pseudo net graph $L_r(2r)$ for $r = 3^m \cdot 2^{m-1}$ follows from Theorem (7.6.1) by choosing $r_1 = 3^{m-1} \cdot 2^{m-2}$, $r_2 = 3$. Hence by induction a pseudo net graph $L_r(2r)$ exists for $r = 3^m \cdot 2^{m-1}$. Thus the theorem holds for $m \geqslant 1$, $n = 0$.

Finally the existence of the pseudo net graph $L_r(2r)$ for $r = 3^m \cdot 2^{m+n-1}$ follows from Theorem (7.6.1) by choosing $r_1 = 3^m \cdot 2^{m-1}$ and $r_2 = 2^{n-1}$, where $m > 1$, $n > 1$. Hence the theorem holds for $m \geqslant 1$, $n \geqslant 1$.

Corollary. $G_2(d)$ graphs with parameters $v = 4r^2$, $n_1 = r(2r-1)$, $d = r(r-1)$ exist for all $r = 3^m \cdot 2^{m+n-1}$ where m and n are non-negative integers and $(m, n) \neq (0, 0)$.

Similarly we can prove [Bose and Shrikhande (1970)] the following theorems:

Theorem (7.6.3). The existence of a pseudo net graph $L_{r_1}(2r_1)$ and a negative Latin square graph $NL_{r_2}(2r_2)$ implies the existence of a negative Latin square graph $NL_r(2r)$, where $r = 2r_1 r_2$.

Theorem (7.6.4). A negative Latin square graph $NL_r(2r)$ exists for all $r = 3^m \cdot 2^{m+n-1}$, where m and n are non-negative integers, $(m, n) \neq (0, 0)$.

Corollary. $G_2(d)$ graphs with parameters $v = 4r^2$, $n_1 = r(2r+1)$, $d = r(r+1)$ exist for all $r = 3^m \cdot 2^{m+n-1}$ where m and n are non-negative integers and $(m, n) \neq (0, 0)$.

7. <u>Descendant of a strongly regular graph</u>. Let G be a strongly regular graph with parameters

$$v, \quad n_1, \quad p_{11}^1, \quad p_{11}^2$$

and vertex set

$$x_0, \quad x_1, \quad \ldots, \quad x_{n_1}, \quad x_{n_1+1}, \quad \ldots, \quad x_{v-1} \cdot$$

where $V_0 = \{x_1, \ldots, x_{n_1}\}$ and $V_2 = \{x_{n_1+1}, \ldots, x_{v-1}\}$ are the sets of vertices adjacent and non-adjacent to x_0. If G_0 is the subgraph of G which remains by deleting x_0 and the edges incident with it, and G_* is obtained from G_0 by complementation with respect to V_1 and V_2 then G_* is defined to be the descendent of G with respect to the vertex x. Bose and Shrikhande (1971) have proved

<u>Theorem (7.7.1)</u>. If G is a strongly regular graph with parameters $v, n_1, p_{11}^1, p_{11}^2$, the necessary and sufficient condition for the descendant G_* of G (with respect to any vertex x_0) to be strongly regular is

$$(7.7.1) \qquad p_{11}^1 + p_{11}^2 = 2n_1 - \frac{v}{2} \cdot$$

When this condition is satisfied the parameters of G_* are given by

$$(7.7.2) \qquad v_* = v - 1, \quad n_{1*} = 2n_1 - 2p_{11}^2 ,$$

$$(7.7.3) \qquad p_{11*}^1 = n_{1*} - n_1 + p_{11}^1, \quad p_{11*}^2 = n_{1*} - n_1 + p_{11}^2 \cdot$$

8. <u>$G_2(d)$ graphs derivable as descendants</u>. Consider a strongly regular graph G with parameters $v, n_1, p_{11}^1, p_{11}^2$, for which the condition (7.7.1) is satisfied. Then its descendant G_* has the parameters (7.7.2) and (7.7.3). If G_* is a $G_2(d_*)$ graph, $p_{11*}^1 = p_{11*}^2$. Hence $p_{11}^1 = p_{11}^2$. Thus G itself must be a $G_2(d)$ graph, for which $p_{11}^1 = p_{11}^2 = d$. Substitut-

ing in (7.7.1) we have $v = 4(n_1 - d)$. It follows that G must either be a pseudo net $L_r(2r)$ graph or a negative Latin square $L_r(2r)$ graph. We are therefore lead to studying descendants of such graphs.

Theorem (7.8.1). The descendant of a pseduo net graph $L_r(2r)$ or a negative Latin square graph $L_r(2r)$ is a pseudo linked block graph $LB_r(2r+1)$.

A pseudo linked block graph $LB_r(2r+1)$ exists for all $r = 3^m \cdot 2^{m+m-1}$ where m and n are non-negative integers $(m, n) \neq (0, 0)$.

Let the parameters of a pseudo net graph $L_r(2r)$ be $v = 4r^2$, $n_1 = r(2r-1)$, $p_{11}^1 = p_{11}^2 = r(r-1)$. The condition (7.7.1) is satisfied and the parameters of the descendant graph G_* given by (7.7.2) are

$$(7.8.2) \qquad v_* = 4r^2 - 1, \quad n_{1*} = 2r^2, \quad p_{11*}^1 = p_{11*}^2 = r^2 .$$

Thus G_* is by definition a pseudo linked block graph $LB_r(2r+1)$.

It can be proved exactly in the same way that the descendant of a negative Latin square graph $L_r(2r)$, has precisely the parameters (7.8.2).

Corollary. $G_2(d)$ graphs with parameters $v = 4r^2 - 1$, $n_1 = 2r^2$, $d = r^2$ exist for all $r = 3^m \cdot 2^{m+n-1}$ where m and n are non=negative integers $(m, n) \neq (0, 0)$.

9. Ascendent of a strongly regular graph. Let G be a strongly regular graph with parameters $(v, n_1, p_{11}^1, p_{11}^2)$. Let (V_1, V_2) be a partition of the vertex set V of G where V_1 and V_2 respectively contain n_1^* and $v - n_1^*$ vertices. Let ∞ be a vertex not in V and let G^* be a graph with vertex set (∞, V). We define adjacency in G^* as follows: The vertex ∞ is adjacent only to vertices of V_1 (and to all vertices of V_1). If x, y are in V, then they are adjacent in G^* if and only if they are adjacent in G and belong both to V_1 or both to V_2, or if they

are nonadjacent in G and belong one to V_1 and the other to V_2. If the graph G^* is strongly regular with parameters $(v^*, n_1^*, p_{11}^{1*}, p_{11}^{2*})$ where v^* is necessarily $v + 1$, then G^* is said to be an ascendant of G.

Bose and Shrikhande (1971) have derived the conditions under which a graph G with parameters $(v, n_1, p_{11}^1, p_{11}^2)$ has an ascendent G^* with parameters $(v^*, n_1^*, p_{11}^{1*}, p_{11}^{2*})$. They prove

Theorem (7.9.1). Let G be a strongly regular graph with parameters $(v, n_1, p_{11}^1, p_{11}^2)$. Then G has an ascendent G^* with parameters $(v^*, n_1^*, p_{11}^{1*}, p_{11}^{2*})$ if and only if the following parametric and structural conditions (P) and (S) are satisfied in G.

(P) $v = 6p_{11}^2 - 2p_{11}^1 - 1$.

(S) The equation

$$x^2 + x(p_{11}^1 - 5p_{11}^2) + vp_{11}^2 = 0$$

has an integral solution n_1^* and there exists a partition (V_1, V_2) of the vertex set V of G with n_1^* vertices in V_1 and $v - n_1^*$ vertices in V_2 such that every vertex in V_1 has $n_1^* - n_1 + p_{11}^1$ adjacent vertices in V_1 and every vertex in V_2 has p_{11}^2 adjacent vertices in V_2.

The parameters of G^* are then given by

$$v^* = v + 1, \quad n_1^*, \quad p_{11}^{1*} = n_1^* - n_1 + p_{11}^1, \quad p_{11}^{2*} = n_1^* - n_1 + p_{11}^2 .$$

It is obvious that any two blocks of a BIBD with $\lambda = 1$ have at most one treatment in common. Consider a BIBD with $r = 2k + 1$, $\lambda = 1$. Then the values of v and b are given by $v = 2k^2 - k$ and $b = 4k^2 - 1$. Consider the blocks as vertices of a graph G and define two blocks as adjacent or nonadjacent according as they have a treatment in common or not. Then [2] G is strongly regular with parameters $(4k^2 - 1, 2k^2, k^2, k^2)$ and satisfies the condition (P) of the above theorem. Also the equation

$f(x) = 0$ has integral solutions $k(2k-1)$ and $k(2k+1)$. Take $n_1^* = k(2k-1)$.
If the $4k^2 - 1$ blocks can be partitioned into sets V_1 and V_2 of $k(2k-1)$
and $(k+1)(2k-1)$ blocks respectively such that each block in V_1 is adjacent to $k^2 - k$ blocks in V_1 and each block in V_2 is adjacent to k^2
blocks in V_2, then the condition (S) is also satisfied. From this
it follows that the set V_1 (respectively V_2) contains each of the $2k^2 - k$
treatments exactly k (respectively $k+1$) times.

We note that a BIBD with $r = 2k+1$, $\lambda = 1$ is a partial geometry
$(r, k, t) = (2k+1, k, k)$. The graph G is then the graph of the dual
configuration and is also a partial geometry $(k, 2k+1, k)$. We can,
therefore, state the following theorem.

Theorem (7.9.2). Let G be the graph of the dual of a BIBD with
$r = 2k+1$, $\lambda = 1$. Then G has an ascendant G* which is a pseudo $L_k(2k)$
graph if and only if the $4k^2 - 1$ blocks of the BIBD can be partitioned
into sets V_1 and V_2 of $k(2k-1)$ and $(k+1)(2k-1)$ blocks respectively
such that each of the $2k^2 - k$ treatments of the BIBD occur k times in V_1
and $k+1$ times in V_2.

BIB designs having the structure of the above theorem exist for
$k = 5$ and 7 [Hall (1967), Appendix I]. Hence we have the following result

Corollary. Pseudo $L_5(10)$ and pseudo $L_7(14)$ graphs exist.

Goethals and Seidel (1970) have constructed a pseudo $L_5(10)$ graph
in precisely the same manner.

Using Theorems (7.6.1), (7.6.2), (7.6.3) and their corollaries we
have the following theorems

Theorem (7.9.3). Pseudo $L_r(2r)$ graphs exist for all
$r = 3^m 5^a 7^c 2^{m+a+c+n-1}$ where m, n, a, c are non-negative integers

$(m, n, a, c) \neq (0, 0, 0, 0)$.

Theorem (7.9.4). $NL_r(2r)$ graphs exist for $r = 5^a 7^c 2^{a+c}$ where, a, c are non-negative integers and for $r = 3^m 5^a 7^c 2^{m+a+c+n-1}$ where m, n, a, c are non-negative and $(m, n) \neq (0, 0)$.

Finally, noting that pseduo $L_r(2r)$ and $NL_r(2r)$ graphs are $G_2(d)$ graphs we have

Theorem (7.9.5). $G_2(d)$ graphs with the following parameters exist

(i) $v = 4r^2$, $n_1 = r(2r-1)$, $d = r(r-1)$;

(ii) $v = 4r^2 - 1$, $n_1 = 2r^2$, $d = r^2$;

for all $r = 3^m 5^a 7^c 2^{m+a+c+n-1}$ where m, n, a, c are non-negative integers $(m, n, a, c) \neq (0, 0, 0, 0)$.

(iii) $v = 4r^2$, $n_1 = r(2r+1)$, $d = r(r+1)$;

with $r = 5^a 7^c 2^{a+c}$, where a, c are non-negative integers and with $r = 3^m 5^a 7^c 2^{m+a+c+n-1}$ where m, n, a, c are non-negative integers and $(m, n) \neq (0, 0)$.

We remark that since our construction is essentially by a composition method, any new $G_2(d)$ graph with parameters as in corollaries to Theorems (7.6.2), (7.6.4) or (7.8.1) can be utilized in conjunction with the above theorem to enlarge such a family considerably.

CHAPTER VIII

GEOMETRIC AND PSEUDO GEOMETRIC GRAPHS $(q^2+1, q+1, 1)$

1. The design corresponding to a vertex of a strongly regular graph.

Let A be the adjacency matrix of a strongly regular graph with parameters

v, n_1, p_{11}^1, p_{11}^2. Noting that A is the association matrix B_1 of the corresponding association scheme, we have from (3.4.5)

(8.1.1) $$A^2 = n_1 I_v + p_{11}^1 A + p_{11}^2 (J_v - I_v - A)$$

where I_v is unit matrix of order v and J_v is a $v \times v$ matrix with each element unity.

Let θ_0 be a particular vertex of G. Let G be the $n_1 \times n_2$ submatrix of A, whose rows correspond to the vertices $\theta_1, \theta_2, \ldots, \theta_{n_1}$ which are adjacent to θ_0, and whose columns correspond to the vertices $\beta_1, \beta_2, \ldots,$ β_{n_2} which are non-adjacent to θ_0. There are exactly p_{12}^1 unities in the row of B corresponding to θ_1, since there are p_{12}^1 vertices among $\beta_1, \beta_2,$ $\ldots, \beta_{n_2}$ which are adjacent to both θ_0 and θ_1. Similarly the column of B which corresponds to β_j has exactly p_{11}^2 unities, since there are p_{11}^2 vertices among $\theta_1, \theta_2, \ldots, \theta_{n_1}$ which are adjacent to both θ_0 and β_j. Hence B is the incidence matrix of a design with parameters $v' = n_1$, $b' = n_2$, $r' = p_{12}^1$, $k' = p_{11}^2$. We shall say that this design corresponds to the vertex θ_0, and denote it by $\mathcal{D}(\theta_0)$.

2. **The properties (P) and (P*).** Let G be a pseudo geometric graph $(r, k, 1)$. Then G will be said to have the property (P) with respect to the vertex θ_0, if the $r(k-1)$ vertices adjacent to θ_0 can be partitioned into r disjoint sets $S_1, S_2, \ldots, S_r$ of size $k-1$ such that any two vertices belonging to the same set S_i are adjacent. Then $K_i = S_i \cup \theta_0$ is a clique. Since $t = 1$, any vertex θ_{iu} which belongs to S_i is non-adjacent to any vertex $\theta'_{i'u'}$ which belongs to $S_{i'}$, $i \neq i'$.

We shall say that G has the additional property (P*) if any two vertices β_j, $\beta_{j'}$ both non-adjacent to θ_0, and both adjacent to θ_{iu} and $\theta_{i'u'}$ are themselves non-adjacent, where θ_{iu} and $\theta_{i'u'}$ are any two vertices belonging to S_i and $S_{i'}$, respectively, $i \neq i'$.

3. <u>The design $\mathcal{D}(\theta_0)$ corresponding to a vertex θ_0 of a pseudo geometric graph $(q^2 + 1, q + 1, 1)$.</u> Let $\mathcal{D}(\theta_0)$ be the design corresponding to the vertex θ_0 if a pseudo-geometric $(q^2 + 1, q + 1, 1)$ graph G, which has the property (P) with respect to the vertex θ_0. The parameters of G are

(8.3.1) $\qquad v = (q + 1)(q^3 + 1), \quad n_1 = q(q^2 + 1), \quad n_2 = q^4 ;$

(8.3.2) $\quad (p^1_{jk}) = \begin{pmatrix} q-1 & q^3 \\ \cdots & q^3(q-1) \end{pmatrix}, \quad (p^2_{jk}) = \begin{pmatrix} q^2+1 & (q^2+1)(q-1) \\ \cdots & q(q-1)(q^2+1) \end{pmatrix} .$

Be definition the treatments of $\mathcal{D}(\theta_0)$ correspond to the vertices of G adjacent to θ_0, and the blocks of $\mathcal{D}(\theta_0)$ correspond to the vertices of G non-adjacent to θ_0. Hence the number of treatments is $v' = n_1 = q(q^2 + 1)$, and the number of blocks is $b' = n_2 = q^4$.

Using the property (P) Bose and Shrikhande (1972) have shown that $\mathcal{D}(\theta_0)$ is group divisible (GD) design with parameters

(8.3.3)
$$v' = q(q^2 + 1), \quad b' = q^4, \quad r' = q^3, \quad k' = q^2 + 1,$$
$$m' = q^2 + 1, \quad n' = q, \quad \lambda'_1 = 0, \quad \lambda'_2 = q^2 .$$

Since $\lambda'_2 v' - r'k' = 0$, the design is semi-regular. Hence each block contains exactly one treatment from each set. We therefore have

<u>Theorem (8.3.1)</u>. If G is a pseudo-geometric graph $(q^2 + 1, q + 1, 1)$ having the property (P) with respect to the vertex θ_0, then the design $\mathcal{D}(\theta_0)$ corresponding to the vertex θ_0 is a semi-regular group divisible (SRGD) design, with parameters (8.3.3).

<u>Corollary</u>. Each vertex non-adjacent to θ_0, is adjacent to exactly one vertex in each of the sets $S_1, S_2, \ldots, S_{q^2+1}$. Two blocks of $\mathcal{D}(\theta_0)$ will be called <u>first associates</u> if they correspond to two vertices in B which are adjacent, and <u>second associates</u> if they correspond to two ver-

tices in B which are non-adjacent. Since each vertex in B is adjacent to p_{12}^2 other vertices in B, and non-adjacent to p_{22}^2 other vertices in B, each block of $\mathcal{D}(\theta_0)$ is first associate of m_1' blocks and second associate of m_2' blocks where

$$(8.3.4) \qquad m_1' = (q^2 + 1)(q - 1), \quad m_2' = q(q - 1)(q^2 + 1) .$$

Under the hypothesis of Theorem (8.3.1) we can now further prove [Bose and Shrikhande (1972)] the following

Corollary. If a treatment occurs in a block β_j of $\mathcal{D}(\theta_0)$, then it occurs $q - 1$ times among the blocks which are first associates of β_j, and $q^3 - q$ times among the blocks which are second associates of β_j. If a treatment does not occur in a block β_j of $\mathcal{D}(\theta_0)$ then it occurs q^2 times among the blocks which are first associates of β_j and $q^3 - q^2$ times among the blocks which are second associates of β_j.

Let us now assume that G has the additional property (P*). We shall show that under this hypothesis any two blocks of $\mathcal{D}(\theta_0)$ intersect in one treatment if they are first associates and in $q + 1$ treatments if they are second associates. Let the block β_j contain two treatments θ_{iu} and $\theta_{i'u'}$. They must belong to different sets. Hence the vertex θ_{iu} belongs to S_i and the vertex $\theta_{i'u'}$ belongs to $S_{i'}$, $i \neq i'$. If the pair θ_{iu}, $\theta_{i'u'}$ occurs in a block $\beta_{j'}$ which is a first associate of β_j, then property (P*) is contradicted. Hence two blocks of $\mathcal{D}(\theta_0)$ which are first associates cannot have more than one treatment in common. Since each treatment of β_j occurs $q - 1$ times among the first associates of β_j, the number of blocks which are first associates of β_j and contain exactly one treatment of β_j is $(q^2 + 1)(q - 1)$. But this is the total number of first associates of β_j. Hence any two blocks of $\mathcal{D}(\theta_0)$ which are first associates intersect in exactly one treatment.

Let us now consider the distribution of the $k' = q^2 + 1$ treatments belonging to the block β_j of $\mathcal{D}(\theta_0)$ among the m_2' blocks which are second associates of β_j, where m_2' is given by (8.3.4). Let x_i be the number of treatments in which β_j intersects the i-th block which is a second associate of β_j, $i = 1, 2, \ldots, m'$. By the second corollary to Theorem (8.3.1) each treatment of β_j occurs $q^3 - q$ among the second associates of β_j. Again any pair of treatments belonging to β_j must occur $\lambda_2' - 1 = q^2 - 1$ times among the second associates of β_j, since it cannot occur among the first associates. Hence

$$\sum_{i=1}^{m_2'} x_i = (q^2 + 1)(q^3 - q) ,$$

$$\sum_{i=1}^{m_2'} x_i(x_i - 1) = (q^2 + 1) q^2 (q^2 - 1) .$$

Let

$$\bar{x} = \sum_{i=1}^{m_2'} x_i/m_2' = q + 1 .$$

Hence

$$\sum_{i=1}^{m_2'} (x_i - \bar{x})^2 = 0 .$$

This shows that $x_i - \bar{x} = 0$, i.e., any two blocks of $\mathcal{D}(\theta_0)$ which are second associates, intersect in exactly $q + 1$ treatments.

Theorem (8.3.2). If the pseudo geometric graph $(q^2 + 1, q + 1, 1)$ of Theorem (8.3.1) has the additional property (P*) with respect to the vertex θ_0, then the design $\mathcal{D}(\theta_0)$ corresponding to the vertex θ_0 has the property (I_1) that any two blocks which are first associates intersect in exactly one treatment and any two blocks which are second associates intersect in $q+1$ treatments.

We can now further prove [Bose and Shrikhande (1972)] the following

Theorem (8.3.3). If G is a pseudo geometric graph $(q^2+1, q+1, 1)$ having the properties (P) and (P*) with respect to the vertex θ_0, then the subgraph G_2 of G, whose vertex set is the set of those vertices of G which are non-adjacent to θ_0, is a strongly regular graph with parameters

(8.3.5) $v^* = q^4$, $n_1^* = (q-1)(q^2+1)$, $p_{11}^{1*} = q-2$, $p_{11}^{2*} = q(q-1)$,

i.e., a negative Latin square graph $NL_{q-1}(q^2)$.

Corollary. The dual $\mathcal{D}^*(\theta_0)$ of $\mathcal{D}(\theta_0)$ is a partially balanced incomplete block (PBIB) design based on a negative Latin square association scheme $NL_{q-1}(q^2)$.

4. Semiregular group divisible designs (SRGD) with the property (I_1). The converse of Theorem (8.3.3) is of great interest. Given an SRGD design $\mathcal{D}$ with parameters (8.3.3), and having the additional property (I_1) that any two blocks of $\mathcal{D}$ intersect in either 1 or $q+1$ treatments, we can ask whether there exists a pseudo geometric graph G having the properties (P) and (P*), with respect to a vertex θ_0 of G such that $\mathcal{D}$ is isomorphic to $\mathcal{D}(\theta_0)$. The answer is in the affirmative as has been proved by Bose and Shrikhande (1972), but as the proof is quite long we shall merely state the theorem obtained by them.

Theorem (8.4.1). Given an SRGD design $\mathcal{D}$ with parameters (8.3.3), having the property (I_1) that any two blocks of $\mathcal{D}$ intersect in either 1 or $q+1$ treatments, there exists a pseudo geometric $(q^2+1, q+1, 1)$ graph G, and a vertex θ_0 of G such that $\mathcal{D}$ is isomorphic with the design $\mathcal{D}(\theta_0)$ corresponding to θ_0, and G has the properties (P) and (P*) with respect to θ_0.

Corollary. If a treatment occurs in any block β_j of $\mathcal{D}$, then it occurs $q - 1$ times among the first associates of β_j and $q^3 - q$ times among the second associates of β_j. If a treatment does not occur in β_j, then it occurs q^2 times among the first associates of β_j and $q^3 - q^2$ times among the second associates of β_j.

5. Partial geometries and geometric graphs $(q^2 + 1, q + 1, 1)$. Let P be a partial geometry $(q^2 + 1, q + 1, 1)$. The points and line of P satisfy axioms A1 - A4 of Chapter IV, paragraph 3 with $r = q^2 + 1$, $k = q + 1$, $t = 1$. Let G be the graph of the geometry, then G is strongly regular with parameters given by (8.3.1) and (8.3.2). Two points of the geometry which are incident with the same line and are therefore adjacent in G, may be called adjacent points. Similarly two points of the geometry may be said to be non-adjacent if there is no line of the geometry incident with both. They are non-adjacent in G.

Let θ_0 be a vertex of G, i.e., a point of P. Let $\ell_i (i = 1, 2, \ldots, q^2 + 1)$ be the lines of P incident with θ_0. Let S_i denote the set of q points (other than θ_0) incident with ℓ_i. Then S the set of points adjacent to θ_0, is the union of the disjoint sets $S_1, S_2, \ldots, S_{q^2+1}$ and G obviously has the property (P) with respect to θ_0. Any two vertices belonging to the same set S_i are adjacent, but if $i \neq i'$, then a vertex belonging to S_i is non-adjacent to a vertex belonging to $S_{i'}$.

It follows from $t = 1$ that there cannot exist a triangle in the geometry, i.e., if θ_1, θ_2, θ_3 are any three distinct points which are pairwise adjacent, then they must be incident with the same line ℓ. From this it easily follows that G has the additional property (P*) with respect to θ_0. It now follows from Theorems (8.3.1) and (8.3.2) that the design $\mathcal{D}(\theta_0)$ corresponding to the vertex θ_0 is an SRGD design with parameters (8.3.3), and possesses the property (I_1) that any two blocks of $\mathcal{D}(\theta_0)$ intersect in either 1 or $q+1$ treatments. Also from the corollary to Theorem (8.4.1), if θ_{iu} is any treatment in a block β_j of $\mathcal{D}(\theta_0)$, then there are exactly $q-1$ other blocks which are first associates of β_j and contain the treatmen θ_{iu}. We can then show that in the present case $\mathcal{D}(\theta_0)$ has the additional property (I_2), that these $q-1$ blocks are mutually first associates. Hence we have the following theorem:

Theorem (8.5.1). If G is the graph of a partial geometry $(q^2+1,\ q+1,\ 1)$, then the design $\mathcal{D}(\theta_0)$ corresponding to any vertex θ_0 is SRGD with parameters (8.3.3), and possesses the properties (I_1) and (I_2).

Consider any three point θ_0, θ_1, θ_2 of the partial geometry $(q^2+1,\ q+1,\ 1)$ which are pairwise non-adjacent. Then θ_1 and θ_2 can be identified with block of $\mathcal{D}(\theta_0)$, which are second associates and therefore intersect in $q+1$ treatments ϕ_0, ϕ_1, $\ldots$, ϕ_q. Since the treatments in any block belong to different sets, ϕ_i and ϕ_j are non-adjacent in G. We therefore have

Corollary. Given any three points θ_0, θ_1, θ_2 of a partial geometry $(q^2+1,\ q+1,\ 1)$ which are pairwise non-adjacent, we can find a set of $q+1$ points ϕ_0, ϕ_1, $\ldots$, ϕ_q which are pariwise non-adjacent and each of which is adjacent to θ_0, θ_1 and θ_2.

We now consider the converse of Theorem (8.5.1). Suppose there exists an SRGD design $\mathcal{D}$ with parameters (8.3.3) and possessing the properties (I_1) and (I_2). Then from Theorem (8.4.1) we can first construct a strongly regular graph G with parameters given by (8.3.1) and (8.3.2), and possessing a vertex θ_0 with respect to which G has the properties (P) and (P*), and such that $\mathcal{D} = \mathcal{D}(\theta_0)$. G is pseudo geometric $(q^2 + 1, q + 1, 1)$. We can now show that in consequence of the additional property (I_2), G must be geometric.

Let $P(V, L, I)$ be an incidence structure consisting of a set of points V, a set of lines L and an incidence relation I defined as follows:

The points of V are the vertices of G. The point corresponding to θ_0 will be said to be of type 0, the points corresponding to the vertices adjacent to θ_0, i.e., to the treatments of $\mathcal{D}(\theta_0)$ will be said to be of type 1, and the points corresponding to the vertices of G non-adjacent to θ_0, i.e., to the blocks of $\mathcal{D}(\theta_0)$ will be said to be type 2. Thus there are v points, of which one is of type 0, n_1 are of type 1, and n_2 are of type 2, where v, n_1, n_2 are given by (8.3.1).

The lines of L are certain subsets of V. The lines of type 1 are the subsets $K_i = S_i \cup \theta_0$, $i = 1, 2, \ldots, q^2 + 1$. Again let θ_{iu} be any treatment of $\mathcal{D}(\theta_0)$, i.e., a point of type 1. Let $\beta_j = \beta_{j0}$ be a block containing θ_{iu}, then θ_{iu} is contained in $q - 1$ first associates of β_j which we have denoted by $\beta_{j1}, \beta_{j2}, \ldots, \beta_{j,q-1}$ and which from property (I_2) are mutually first associates. Then we shall call the set of points $\theta_{iu}, \beta_{j0}, \beta_{j1}, \ldots, \beta_{j,q-1}$ a line of type 2. The treatment θ_{iu} occurs in each of the blocks $\beta_{j0}, \beta_{j1}, \ldots, \beta_{j,q-1}$ and is the only treatment common to any two of them. Hence in defining the line we could have started from θ_{iu} and any one of the blocks $\theta \beta_{jw}$, $w = 0, 1, \ldots, q-1$.

The incidence relation I is the containing contained relation. We can then show [Bose and Shrikhande (1972)] that $P(V, L, I)$ satisfies axioms A1 - A4 for a partial geometry. Thus the incidence structure $P(V, L, I)$ is a partial geometry, and G is the graph of this geometry. Thus G is a geometric graph. We thus have

Theorem (8.5.2). Given an SRGD design with parameters (8.3.3) and possessing the properties (I_1) and (I_2), there exists a partial geometry $(q^2+1, q+1, 1)$, such that $D = D(\theta_0)$ where $D(\theta_0)$ is the design corresponding to some vertex of the graph G of the partial geometry.

Now the partial geometry $(q^2+1, q+1, 1)$ is known. We have pointed out in Chapter VII, paragraph 5(a) two ways of obtaining it. Higman has shown (unpublished) that the two realizations are isomorphic. It follows from the theorem above

Corollary. If q is a prime or a prime power, an SRGD design D with parameters (8.3.3) and possessing the properties (I_1) and (I_2) exists. Also an $NL_{q-1}(q^2)$ association scheme or graph exists.

6. Unsolved problems and conjectures. Let θ_0, θ_1, θ_2 be any three pairwise non-adjacent points of a partial geometry $(q^2+1, q+1, 1)$. Then we have shown in corollary to Theorem (8.5.1), that there exists a set of $q+1$ points ϕ_0, ϕ_1, $\ldots$, ϕ_{q^2+1} which are pairwise non-adjacent and each of which is adjacent to θ_0, θ_1, θ_2. Now starting with the non-adjacent points ϕ_0, ϕ_1, ϕ_2 we may likewise obtain a set $q+1$ points θ_0, θ_1, θ_2, $\ldots$, θ_{q^2+1} which are pairwise non-adjacent and each of which is adjacent to ϕ_0, ϕ_1, ϕ_2. We may conjecture that θ_i is adjacent to ϕ_j for all i, j = 1, 2, $\ldots$, q^2+1.

Again one may ask whether a partial geometry $(q^2+1, q+1, 1)$ exists when q is not a prime power. Even if such a geometry does not exist for a given q, a pseudo geometric graph $(q^2+1, q+1, 1)$ might exist.

If q is a prime power one may ask whether all partial geometries $(q^2+1, q+1, 1)$ are isomorphic. If the answer to this question is in the affirmative, then our conjecture stated above is certainly true since it is easily proved by geometrical considerations for the partial

geometry derived from the elliptic quadric Q in PG(5, q). But the conjecture could still be true even if there exist non-isomorphic partial
geometries $(q^2 + 1, q + 1, 1)$.

Again let ℓ_0 and ℓ_1 be two non-intersecting lines of a partial geom-
etry $(q^2 + 1, q + 1, 1)$. Let θ_{00}, θ_{01}, θ_{02}, ..., θ_{0q} be the points of ℓ_0.
Through θ_{0j} there passes a unique line m_j meeting ℓ_1 in a point which we
may denote by θ_{ij} (j = 0, 1, 2, ..., q). The condition t = 1 shows that
the points θ_{10}, θ_{11}, ..., θ_{1q} are all distinct. Let the points of m_0 be
θ_{00}, θ_{10}, ..., θ_{q0}. Through any point of θ_{10} of m_0 there passes a unique
line ℓ_i meeting m_1 in a point which we may denote by
θ_{i1}, (i = 0, 1, 2, ..., q). The points θ_{01}, θ_{11}, ..., θ_{q1} are all distinct. We may conjecture that ℓ_i and m_j intersect in a point
θ_{ij} (i, j = 0, 1, ..., q).

Again this conjecture is certainly true if a partial geometry
$(q^2 + 1, q + 1, 1)$ is always isomorphic to the partial geometry derived
from the elliptic quadric Q (as can be proved by geometrical considerations) but it could still be true even if non-isomorphic partial geometries $(q^2 + 1, q + 1, 1)$ exist.

R. C. BOSE

BIBLIOGRAPHY

1. Aigner, M. (1969). A characterization problem in graph theory. J. Comb. Theory 6, 45 - 55.

2. Archbold, J. W. and Johnson, N. L. (1956). A method of constructing partially balanced incomplete block designs. Ann. Math. Statist. 27, 633 - 641.

3. Bhattacharya, K. N. (1944). A new balanced incomplete block design. Science and Culture 9, 508.

4. Bose, R. C. (1939). On the construction of balanced incomplete block designs. Ann. Eugenics 9, 353 - 399.

5. _____ (1942 a). A note on the resolvability of balanced incomplete block designs. Sankhya 6, 105 - 110.

6. _____ (1942 b). On some new series of balanced incomplete block designs. Bull. Cal. Math. Soc. 34, 17 - 31.

7. _____ (1947 a). Mathematical theory of the symmetrical factorial design. Sankhya 8, 107 - 166.

8. _____ (1947 b). On a resolvable series of balanced incomplete block designs. Sankhya 8, 249 - 256.

9. _____ (1949). A note on Fisher's inequaulity for balanced incomplete block designs. Ann. Math. Statist. 20, 619 - 620.

10. _____ (1952). A note on Nair's condition for partially balanced incomplete block designs with $k > r$. Calcutta Statist. Assoc. Bull. 4, 123 - 126.

11. _____ (1963 a). On the application of finite projective geometry for deriving a certain series of balanced Kirkman arrangements. Calcutta Math. Soc. Golden Jubilee Commem. vol. (1958/59), part II, 341 - 354.

12. _____ (1963 b). Combinatorial properties of partially balanced designs and association schemes. Sankhya 25, 109 - 136.

13. _____ (1963 c). Strongly regular graphs, partial geometries and partially balanced designs. Pacific J. Math. 13, 389 - 419.

14. Bose, R. C. and Chakravarti, I. M. (1966). Hermitian varieties in a finite projective space PG (N, q^2). Canad. J. Math. 18, 1161 - 1182.

R. C. BOSE

15. Bose, R. C., Chakravarti, I. M. and Knuth, D. K. (1960). On methods of constructing sets of mutually orthogonal Latin squares using a computer. I. Technometries 2, 507 - 516.

16. Bose, R. C. and Clatworthy, W. H. (1955). Some classes of partially balanced designs. Ann. Math. Statist. 26, 212 - 232.

17. Bose, R. C., Clatworthy, W. H. and Shrikhande, S. S. (1954). Tables of partially balanced designs with two associate classes. N. C. Ag. Expt. Station Tech. Bull. 107, Raleigh, N. C.

18. Bose, R. C. and Connor, W. S. (1952). Combinatorial properties of group divisible incomplete block designs. Ann. Math. Statist. 23, 367 - 383.

19. Bose, R. C. and Dowling, R. A. (1971). A generalization of Moore graphs of diameter two. J. Comb. Theory 11, 213 - 226.

20. Bose, R. C. and Lasker, R. (1967). A characterization of tetrahedral graphs. J. Comb. Theory 3, 366 - 385.

21. Bose, R. C. and Mesner, D. M. (1959). On linear associative algebras corresponding to the association schemes of partially balanced designs. Ann. Math. Statist. 30, 21 - 38.

22. Bose, R. C. and Nair, K. R. (1939). Partially balanced incomplete block designs. Sankhya 4, 337 - 372.

23. Bose, R. C. and Shimamoto, T. (1952). Classification and analysis of partially balanced incomplete block designs with two associate classes. J. Amer. Statist. Assoc. 47, 151 - 184.

24. Bose, R. C. and Shrikhande, S. S. (1960 a). On the composition of balanced incomplete block designs. Canad. J. Math. 12, 177 - 188.

25. _____ (1960 b). On the construction of sets of mutually orthogonal Latin squares and the falsity of a conjecture of Euler. Trans. Amer. Math. Soc. 95, 190 - 209.

26. _____ (1970). Graphs in which each pair of vertices is adjacent to the same number d of other vertices. Studia Sci. Math. Hungarica 5, 181 - 195.

27. _____ (1971). Some further constructions for $G_2(d)$ graphs. Studia Sci. Math. Hungarica 6, 127 - 132.

28. _____ (1972). Geometric and pseudo geometric graphs $(q^2+1, q+1, 1)$. J. Geometry 2/1, 75 - 94.

29. _____ (1973). Embedding the complement of an oval in a pro-jective plane of even order. J. of Discrete Math.

30. Bose, R. C., Shrikhande, S. S. and Bhattacharya, K. N. (1953). On the construction of group divisible incomplete block designs. Ann. Math. Statist. 24, 167 - 195.

31. Brauer, A. (1952). Limits for the characteristic roots of a matrix. IV: Applications to stochastic matrices. Duke Math. J. 19, 75 - 91.

32. Bruck, R. H. (1963). Finite nets II. Uniqueness and imbedding. Pacific J. Math. 13, 421 - 457.

33. Bruck, R. H. and Ryser, H. J. (1949). The nonexistence of certain finite projective planes. Canad. J. Math. 1, 88 - 93.

34. Chang Li-Chien (1959). The uniqueness and non-uniqueness of tri-angular association schemes. Science Record, Math. New Ser. 3, 604 - 613.

35. _____ (1960). Assocation schemes of partially balanced design with parameters $v = 28$, $n_1 = 12$, $n_2 = 15$, and $p_{11}^2 = 4$. Science Record, Math. New Ser. 4, 12 - 18.

36. Chowla, S. and Ryser, H. J. (1950). Combinatorial problems. Canad. J. Math. 2, 93 - 99.

37. Clatworthy, W. H. (1954). A geometrical comfiguration which is a partially balanced design. Proc. Amer. Math. Soc. 5, 47 - 55.

38. _____ (1955). Partially balanced incomplete block designs with two associate classes and two treatments per block. J. Res. Nat. Bur. Standards 54, 177 - 190.

39. _____ (1956). Contributions on partially balanced incomplete block designs with two associate classes. Nat. Bur. Standards. App. Math. Ser. No. 47, Washington, D.C.

40. Connor, W. S. (1952). On the structure of balanced incomplete block designs. Ann. Math. Statist. 23, 57 - 71.

41. _____ (1958). The uniqueness of the triangular association scheme. Ann. Math. Statist. 29, 262 - 266.

42. Connor, W. S. and Clatworthy, W. H. (1954). Some theorems for partially balanced designs. Ann. Math. Statist. 25, 100 - 112.

R. C. BOSE

43. Dowling, T. A. (1968). A characterization of cubic lattice graphs. J. Comb. Theory 5, 425 - 426.

44. _____ (1969). A characterization of the T_m graph. J. Comb. Theory 6, 251 - 263.

45. Erdös, P., Renyi, A. and Sós, V. T. (1966). On a problem of graph theory. Studia Sci. Math. Hungarica 1, 215 - 235.

46. Fisher, R. A. (1940). An examination of the different possible solutions of a problem in incomplete blocks. Ann. Eugerics 10, 52 - 75.

47. _____ (1942). New cyclic solutions to problems in incomplete blocks. Ann. Eugenics 11, 290 - 299.

48. Goethals, J. M. and Seidel, J. J. (1970). Strongly regular graphs derived from combinatorial designs. Canad. J. Math. 22, 449 - 471.

49. Hall, M. (1967). Combinatorial theory. Blaisdell, Waltham, Mass.

50. Hall, M. and Connor, W. S. (1953). An embedding theorem for balanced incomplete block designs. Canad. J. Math. 6, 35 - 41.

51. Hanani, H. (1961). The existence and construction of balanced incomplete block designs. Ann. Math. Statist. 32, 361 - 386.

52. _____ (1965). A balanced incomplete block design. Ann. Math. Statist. 36, 711.

53. Hoffman, A. J. (1960 a). On the uniqueness of the triangular association scheme. Ann. Math. Statist. 31, 492 - 497.

54. _____ (1960 b). On the exceptional case in a characterization of the arcs of a complete graph. IBM J. Res. Develop. 4, 497 - 504.

55. Lasker, R. (1967). A characterization of cubic lattice graphs. J. Comb. Theory 3, 386 - 401.

56. Mann, H. B. (1949). Analysis and design of experiments. Dover, New York.

57. Masuyama, M. (1961). Calculas of blocks and a class of partially balanced incomplete block designs. Rep. Statist. Appl. Res. Un. Japan Sci. Engrs. 8, 59 - 69.

58. _____ (1964 a). Construction of PBIB designs by fractional development. Rep. Statist. Appl. Res. Un. Japan Sci. Engrs. 11, 47 - 54.

R. C. BOSE

59. _____ (1964 b). Linear graphs of PBIB designs. Rep. Statist.
Appl. Res. Un. Japan Sci. Engrs. 11, 147 - 151.

60. Mesner, D. M. (1967). A new family of partially balanced designs
with some Latin square design properties. Ann. Math. Statist. 38,
571 - 581.

61. Nair, K. R. (1943). Certain inequality relations among the combinatorial parameters of balanced incomplete block designs. Sankhya
6, 255 - 259.

62. _____ (1950). Partially balanced incomplete block designs
involving only two replications. Calcutta Statist. Assoc. Bull. 3,
83 - 86.

63. _____ (1951 a). Some two replicate PBIB designs. Calcutta
Statist. Assoc. Bull. 3, 174 - 176.

64. _____ (1951 b). Some three-replicate PBIB design. Calcutta
Statist. Assoc. Bull. 4, 39 - 42.

65. Ogawa (1959). A necessary condition for existence of regular and
symmetricl experimental designs of triangular type, with partially
balanced incomplete blcoks. Ann. Math. Statist. 30, 1063 - 1071.

66. _____ (1960). On a unified method of deriving necessary
conditions for existence of symmetrically balanced incomplete block
designs of certain types. Bull. Inst. Internat. Statist. 38, 43 -
57.

67. Primrose, E. J. F. (1951). Quadratics in finite geometries. Proc.
Camb. Phil. Soc. 47, 299 - 304.

68. Qvist, B. (1952). Some remarks concerning curves of the second
degree in a finite plane. Ann. Acad. Sci. Fenn., No. 134, 1 - 27.

69. Rao, C. R. (1946). Difference sets and combinatorial arrangements
derivable from finite geometries. Proc. Nat. Inst. Sci. India 12,
123 - 135.

70. _____ (1961). A study of BIB designs with replications 11 - 15.
Sankhya 23, 117 - 127.

71. Ray Chaudhuri, D. K. (1962 a). Some results on quadrics in finite
projective geometry. Canad. J. Math. 14, 129 - 138.

72. _____ (1962 b). Application of the geometry of quadrics for
constructing PBIB designs. Ann. Math. Statist. 33, 1175 - 1186.

R. C. BOSE

73. _____ (1965). Some configurations in finite projective
    spaces and partially balanced incomplete block designs. Canad. J.
    Math. 17, 114 - 123.

74. Ray Chaudhuri and Wilson (1971). Solution of Kirkman's school girl
    problem. Combinatorics, Amer. Math. Soc, Symp. Pure Math. 19,
    187 - 203.

75. Schutzenberger, M. P. (1949). A non-existence theorem for an,
    infinite family of symmetrical block designs. Ann. Eugenics 14,
    286 - 287.

76. Segre, B. (1954). Sulle ovali dei piani lineari finiti. R. C.
    Acc. Naz. Lincei 17, 141 - 142.

77. _____ (1955). Ovals in a finite projective plane. Canad. J.
    Math. 7, 414 - 416.

78. Seidel, J. J. (1967). Strongly regular graphs of L_2 type and tri-
    angular type. Koninkl Nederl. Akademie Van Wetenschappen - Amster-
    dam Proceedings series A, 70 and Indag. Math. 29, 188 - 196.

79. Shrikhande, S. S. (1950). The impossibility of certain symmetrical
    balanced incomplete block designs. Ann. Math. Statist. 21, 106 -
    111.

80. _____ (1952). On the dual of certain balanced incomplete
    block designs. Biometrics 8, 66 - 72.

81. _____ (1959 a). On a characterization of the triangular asso-
    ciation scheme. Ann. Math. Statist. 30, 39 - 47.

82. _____ (1959 b). The uniqueness of the L_2 association scheme.
    Ann. Math. Statist. 30, 781 - 798.

83. _____ (1960). Relations between incomplete block designs.
    Contributions to probability and statistics. Essays in honor of
    Harold Hotelling. Stanford U. Press, 388 - 395.

84. _____ (1961). A note on mutually orthogonal Latin squares.
    Sankya 23, 115 - 116.

85. _____ (1962). On a two parameter family of balanced incomplete
    block designs. Sankhya, ser. A, 24, 33 - 40.

86. _____ (1965). On a class of partially balanced incomplete
    block designs. Ann. Math. Statist. 36, 1807 - 1814.

R. C. BOSE

87. Shrikhande, S. S. and Raghavarao, D. (1966). A note on the non-existence of symmetric balanced incomplete block designs. Sankhya 26, 91 - 92.

88. Wilson, R. M. (1972 a). An existence theory for pariwise balanced designs. I. Composition theorems and morphisms. J. Comb. Theory (A) 13, 220 - 245.

89. _____ (1972 b). An existence theory for pairwise balanced designs. II. The structure of PBD closed sets and existence con-jectures. J. Comb. Theory (A) 13, 246 - 273.

CENTRO INTERNAZIONALE MAYEMATICO ESTIVO

(C. I. M. E)

R. H. BRUCK

CONSTRUCTION PROBLEMS IN FINITE PROJECTIVE SPACES

Corso tenuto a Bressanone dal 18 al 27 Giugno 1972

Construction Problems in Finite Projective Spaces

by R. H. Bruck

(University of Wisconsin)

Forward. The present paper represents a complete reworking of eight lectures given in Bressanone, Italy in the summer of 1972. The spirit of the lectures has been preserved (or so I hope) and also the order ⸝of the material. However, a few topics are treated more carefully here than seemed desirable in the lectures, and a few results have been added that were unknown in the summer of 1972. The latter – at least insofar as they were not due to me – should be easy to identify in the paper.

1. Spreads and packings of designs. Intrinsic construction of complements of 1-designs. In a series of unpublished lectures given in the summer of 1963 in Saskatoon, Canada (Bruck [2]) I proposed various methods of constructing finite projective planes of order n having one or more affine or projective subplanes of order not dividing n. Common to these methods was the idea of determining the projective plane in terms of simpler substructures. One of the methods led me to the concepts of spreads and packings of 3-dimensional projective space, and these I studied and generalized during the following year in Chapel Hill, North Carolina, in close collaboration with R. C. Bose and Dale Mesner. Later, in 1967, at a conference in Chapel Hill (Bruck [3]) I crystalized the essence of the Saskatoon proposals in a brief preliminary section.

Now I wish to re-examine these ideas, with the hope of focusing attention on some interesting combinatorial problems.

R. H. Bruck

Let v, b, k, r, λ be positive integers with

(1. 1) $v > 1$.

By a design $D = (\mathcal{V}, \mathcal{B}, I)$ with parameters v, b, k, r, λ we mean a system D consisting of a set $\mathcal{V}$ of v (> 1) distinct objects, called varieties of D, a set $\mathcal{B}$ of b distinct objects, called blocks of D, and an incidence relation, I, a subset of $\mathcal{V} \times \mathcal{B}$. We write xIy, for $x \in \mathcal{V}$, $y \in \mathcal{B}$, and say that x is incident with y (or y is incident with x) if and only if the ordered pair (x, y) is in I. We impose the following axioms: (a) Each block of D is incident with exactly k distinct varieties of D. (b) Each variety of D is incident with exactly r distinct blocks of D. (c) If x, y are distinct varieties of D, there are exactly λ blocks of D incident with both of x, y. Such a design may be called a λ-design. The following relations are easily verified:

(1. 2) $k > 1; \; r(k - 1) = \lambda(v - 1); \; vr = bk$.

By a spread, $\mathcal{S}$, of a λ-design D we mean a collection, $\mathcal{S}$, of blocks of D which partition the varieties of D in the sense that each variety of D is incident with one and only one of the blocks in $\mathcal{S}$.

By a packing, $\mathcal{P}$, of a λ-design D we mean a collection, $\mathcal{P}$, of spreads of D which partition the blocks of D; that is, each block of D is in one and only one of the spreads in $\mathcal{P}$.

If a λ-design D possesses a spread $\mathcal{S}$, then

(1. 3) k divides v .

Indeed, if $\mathcal{S}$ contains exactly t distinct blocks, then $v = kt$. In partic- ular, if (1. 3) holds, then a spread of D is merely a set of $t = v/k$ dis-

tinct blocks of D which are disjoint in the sense that no variety of D is incident with more than one of them. Furthermore, since, by the last equation of (1. 2),

$$b = vr/k = tr ,$$

a packing of D is merely a set of r disjoint spreads of D.

As the literature shows, there are many applications of spreads and (to a lesser extent) of packings. We shall be more explicit later. For the time-being, we wish to restrict attention to spreads and packings of 1-designs $(\lambda = 1)$.

Let D be a 1-design. Explicitly, assume (1. 1), (1. 2), (1. 3) together with

$$(1. 4) \qquad\qquad \lambda = 1 .$$

From (1. 4) in the first equation of (1. 2), we get

$$rk - v = r - 1 = n$$

where we define $n \geq 0$ by

$$(1. 5) \qquad\qquad r = n + 1.$$

By this and (1. 3),

$$(1. 6) \qquad\qquad n = kk' ,$$

$$(1. 7) \qquad\qquad v/k = r - k' = 1 + k'(k - 1)$$

where k' is a (unique) non-negative integer. From the second equation of (1. 1), and from (1. 7),

R. H. Bruck

(1. 8) $b/r = r - k' = 1 + k'(k - 1)$.

If $k' = 0$, then $n = 0$, $r = 1$, $v = k$, $b = r = 1$. We shall leave aside this trivial case.

If $k' = 1$ then

$$n = k, \ r = k + 1, \ v = k^2, \ b = k^2 + k .$$

If we regard the varieties and blocks of D as points and lines, respectively, then, since $\lambda = 1$, D becomes an affine plane of order k. It is easy to see that D has precisely $k + 1$ distinct spreads, one for every parallel class, and precisely one packing,   consisting of the $k + 1$ spreads. In this case, the spreads serve as "points at infinity" and the packing as the "line at infinity" in the usual imbedding of the affine plane of order k in a projective plane or order k.

For the rest of the discussion, we assume

(1. 9) $k' > 1$.

We notice that equations (1. 5) through (1. 8) allow us to express each of v, b, k, r (and $\lambda = 1$) in terms of k, k' or (as we prefer) in terms of n, k, k'. The result of interchanging k, k' is to give a new set of parameters v', b', k', r' (and $\lambda' = 1$). If we also use the symbol n', then

(1. 10) $r' = r, \ n' = n$

and

(1. 11) $v'/k' = b'/r = r - k = 1 + k(k' - 1)$.

Also, from (1. 7), (1. 8), (1. 11) and (1. 6),

R. H. Bruck

(1. 12) $$v + b' = v' + b = n^2 + n + 1 \, .$$

Now let us regard the varieties and blocks of D as the points and lines, respectively, of a partial plane π_0, (with incidence given by the incidence relation of D). We ask under what conditions π_0 can be imbedded in a projective plane π subject to the following requirement:

(*)   The only lines of π incident with a point P of π_0 are the lines of π_0 incident (in π_0) with P.

If such a projective plane π exists, then we know its order. Indeed, every point P of π_0 is incident with precisely $r = n + 1$ lines of π; hence π has order n. Moreover, since π_0 has v distinct points, b distinct lines, whereas π has $n^2 + n + 1$ distinct points and a like number of distinct lines, then, by (1. 12), the numbers of distinct lines and points (note the order!) of π which are not in π_0 are v' and b' respectively. Consider a point Q of π which is not π_0. If a line L of π is incident with Q and also with a point of π_0, then, by (*), L must be a line of π_0. Therefore L is incident with exactly k distinct points of π_0. Consequently, Q must be incident (in π) with precisely

$$v/k = r - k' = n + 1 - k'$$

distinct lines of π_0 - and, clearly, these lines form a spread of π_0 (or D). Since Q is incident in π with precisely $n + 1$ lines of π, then Q must be incident in π with precisely k' distinct lines of π which are not in π_0. Next let M be a line of π which is not in π_0. Then, by (*), no point of π_0 is incident (in π) with M. Hence M is incident in π with $n + 1 = r$ distinct points of π which are not in π_0.

Furthermore, the r spreads of π_0 defined by the r points of M partition the lines of π_0 - since every line of π_0 must "meet" M in a point - and hence form a packing of π_0.

Now (assuming that π exists subject to (*)) let D' be the system of which the varieties are the lines of π not in π_0, the blocks are the points of π not in π_0, and the incidence relation is that induced by the incidence relation of π. Then, by the above discussion, D' is a design with parameters v',b',k',r',λ' (with r' = r, λ' = 1) where v',b', are defined by (1.11). We shall call such a design D' a <u>complement</u> of D.

Note that we do not accept every design with the correct parameters as a complement of D. The design D and complement D', by definition, fit together as indicated above to define a projective plane of order n.

Note further that, if π exists subject to (*), then two disjoint blocks B,C of D, considered as lines of π_0, must be incident in π with a unique point Q of π not in π_0, and hence must be contained in the unique spread of D defined by Q. With this as a clue, we can see how to construct a complement D' of D (if one exists) in terms of spreads and packings of D.

We need the existence of a collection, $\mathcal{C}_s$ of spreads of D and a collection, $\mathcal{C}_p$, of packings of D subject to the following requirements:

(I)   Each packing in $\mathcal{C}_p$ is made up of spreads in $\mathcal{C}_s$.

(II)   If B,C are two disjoint blocks of D, there exists one and only one spread in $\mathcal{C}_s$ which contains B and C.

(III)   If $\mathcal{S}_1, \mathcal{S}_2$ are two disjoint spreads in $\mathcal{C}_s$, there exists one and only one packing in $\mathcal{C}_p$ which contains $\mathcal{S}_1$ and $\mathcal{S}_2$.

Now let D' be the system of which the varieties are the packings in $\mathcal{C}_p$, the blocks are the spreads in $\mathcal{C}_s$, and the incidence relation

R. H. Bruck

is the containing relation. If $\mathcal{P}, \mathcal{S}$ are in $\mathcal{C}_p, \mathcal{C}_s$ respectively, then $\mathcal{P}$ is incident with $\mathcal{S}$ iff $\mathcal{S}$ is contained in $\mathcal{P}$. We claim that D' is a design and a complement of D.

To begin with, we need to show that the cardinal numbers of $\mathcal{C}_p, \mathcal{C}_s$ are given by

(1. 13) $|\mathcal{C}_p| = v'$, $|\mathcal{C}_s| = b'$.

The proof of (1. 13) will show somewhat more.

Consider a block B of D. We can choose a treatment x of D which is not incident with B in

$$v - k = n(k - 1)$$

ways (cf. (1. 7), (1. 6)). If y is a treatment incident with B, there is a unique block B_1 of D incident with x and y; as y ranges over the k treatments incident with B, B_1 ranges over k distinct blocks. Hence the number of distinct blocks of D which are incident with x and disjoint from B is (cf. (1. 11))

$$r - k = v'/k' .$$

Therefore the number of ordered pairs x, C such that x is a treatment and C is a block of D, x is incident with C but not with B, and C is disjoint from B is

$$n(k - 1)(v'/k') = k(k - 1)v'$$

(cf. (1. 6)). For given C, x can be chosen in k ways. Hence

(1. 14)   For each block B of D, there are exactly $(k - 1)v'$ distinct blocks of D disjoint from B .

R. H. Bruck

Now we use (II). If C is a block of D disjoint from B, there is a unique spread in $\mathcal{C}_s$ containing B, C, and $\mathcal{S}$ contains exactly (cf. (1.7))

$$(v/k) - 1 = k'(k - 1)$$

distinct blocks disjoint from B. By this and (1.14):

(1.15)　　　　　Each block of D is contained in exactly

$$v'/k' = v'/r \text{ distinct spreads of } D \text{ in } \mathcal{C}_s \text{ .}$$

Since D has b distinct blocks, and since each spread consists of $v/k = b/r$ (disjoint blocks, we see from (1.14) that

(1.16)　　　　　$$|\mathcal{C}_s| = b(v'/r)/(v/r) = b' \text{ .}$$

This proves the second equation of (1.13).

Next consider a spread $\mathcal{S}$ in $\mathcal{C}_s$. For each block B in $\mathcal{S}$, by (1.15), the number of distinct spreads $\mathcal{S}'$ in $\mathcal{C}_s$ which contain B but are distinct from $\mathcal{S}$ is (cf. (1.1))

$$(v'/k') - 1 = k(k' - 1) \text{ .}$$

Each such $\mathcal{S}'$ intersects $\mathcal{S}$ in B, by (II). The number of distinct blocks B in $\mathcal{S}$ is v/k. Hence the number of distinct spreads in $\mathcal{C}_s$ which meet $\mathcal{S}$ in a block is

$$v(k' - 1) \text{ .}$$

Since

$$1 + v(k' - 1) = 1 + k(r - k')(k' - 1)$$

and

R. H. Bruck

$$b' = r + rk(k' - 1) ,$$

then

$$b' - \{1 + v(k' - 1)\} = r - 1 + kk'(k' - 1)$$

$$= n + n(k' - 1) = nk' .$$

Therefore

(1.17)   Each spread in $\mathcal{C}_s$ is disjoint from exactly
nk' distinct spreads in $\mathcal{C}_s$.

Next we use (III). Given a spread $\mathcal{S}$ in $\mathcal{C}_s$, we can choose a spread $\mathcal{S}'$ in $\mathcal{C}_s$ disjoint from $\mathcal{S}$ in exactly nk' ways, by (1.17). By (III), $\mathcal{S}, \mathcal{S}'$ are contained in a unique packing, $\mathcal{P}$, in $\mathcal{C}_p$. The number of distinct spreads in $\mathcal{P}$ distinct from (hence disjoint from) $\mathcal{S}$ is $r - 1 = n$. Hence

(1.18)   Each spread in $\mathcal{C}_s$ is contained in exactly
k' distinct packings in $\mathcal{C}_p$.

Since, by (1.18),

$$|\mathcal{C}_p| = |\mathcal{C}_s| \cdot k'/r = (b'/r)k' = (v'/k')k' = v' ,$$

we have completed the proof of (1.13).

We have a little more to do. Consider a packing $\mathcal{P}$ in $\mathcal{C}_p$. Then $\mathcal{P}$ consists of $r = n + 1$ distinct and disjoint spreads of D. By (I), each of these is in $\mathcal{C}_s$. For each spread $\mathcal{S}$ of $\mathcal{P}$, there are, by (1.18), exactly $k' - 1$ distinct packings in $\mathcal{C}_p$ which contain $\mathcal{S}$ and are distinct from $\mathcal{P}$. If $\mathcal{P}'$ is such a packing, $\mathcal{P} \cap \mathcal{P}' = \mathcal{S}$ by (III).

R. H. Bruck

Hence the number of distinct packings in $\mathcal{C}_p$ which intersect $\mathcal{P}$ in a (single) spread is

$$(n + 1)(k' - 1) = k' - 1 + n(k' - 1)$$

$$= -1 + k'[1 + k(k' - 1)]$$

$$= -1 + v'.$$

In other words:

(1. 19)   Every two distinct packings in $\mathcal{C}_p$
             have a unique common spread in $\mathcal{C}_s$.

In view of (1. 13), (1. 19), (1. 18), we see that D', as defined above in terms of $\mathcal{C}_s$, $\mathcal{C}_p$, is a design with parameters v', b', k', r, λ ($\lambda = 1$). Now we define π as a system whose points are the treatments of D and the blocks of D' and whose lines are the blocks of D and the treatments of D'. Within D or D', the incidence relation of π is that of D or D', respectively. Incidence in π between a block of D' (that is, a spread of D in $\mathcal{C}_p$) and a block of D is defined by the containing relation. Finally, by fiat, no treatment of D' (that is, no packing of D in $\mathcal{C}_p$) is incident in π with a treatment of D. Then π is a projective plane of order n, and D' is a complement of D.

By symmetry, D is a complement of D'.

In the next few sections, we shall discuss 1-designs D with parameters of form (1. 1) - (1. 9), for special values of k.

R. H. Bruck

2. (k = 2.) Cliques. Complete symmetric graphs. Complete
ovals. Let D be a 1-design corresponding to the situation of section 1
in which

(2. 1) $\qquad k = 2 , \ k' = m > 1 ,$

where m is an integer. Then the parameters of D are given by

(2. 2) $\quad v = 2m + 2 , \ b = (m + 1)(2m + 1) , \ k = 2 , \ r = 2m + 1 , \ \lambda = 1 .$

Also

(2. 3) $\qquad n = 2m .$

Furthermore, the parameters of a complement D' of D (if one exists) are
given by

(2. 4) $\quad v' = m(2m - 1) , \ b' = 4m^2 - 1 , \ k' = m , \ r' = 2m + 1 , \ \lambda' = 1 .$

The design D is variously called a complete symmetric graph (or
a clique) with v vertices, a complete v-point or a complete (or completed)
oval. In this special case, the concept of a complement seems to have
been introduced by Esther Seiden [13], and I was aware of some of her work
before formulating the material of §1.

Let us review the concept of an oval, $\mathcal{O}$, of a projective plane π
of finite order n. By definition, $\mathcal{O}$ is a set of n + 1 distinct points of
π no three of which lie on a line of π. We call a line L of π a secant,
tangent or non-secant to $\mathcal{O}$ according as L contains 2, 1 or 0 distinct points of $\mathcal{O}$. Every point of $\mathcal{O}$ lies on n distinct secants and a
unique tangent.

If n is odd, the points of π not on $\mathcal{O}$ form two classes, the

R. H. Bruck

interior points of $\mathcal{O}$ and the exterior points of $\mathcal{O}$. An interior point lies on $(n + 1)/2$ distinct secants, 0 tangents, and $(n + 1)/2$ distinct non-secants. An exterior point lies on $(n - 1)/2$ distinct secants, 2 distinct tangents, and $(n - 1)/2$ distinct secants. There are $(n^2 - n)/2$ distinct interior points, $(n^2 + n)/2$ distinct exterior points, and the $n + 1$ points of $\mathcal{O}$.

If n is even, assume (2. 3). The points of π not on $\mathcal{O}$ form two classes, one consisting of a single point, K, called the _knot_ of $\mathcal{O}$. The point K lies on each of the $n + 1 = 2m + 1$ tangents and on no other line of π. Each of the remaining $n^2 - 1 = 4m^2 - 1$ points not on $\mathcal{O}$ lies on m distinct secants, a single (tangent) line through K, and m non-secants. We complete $\mathcal{O}$ to $\mathcal{O}^*$ by adjoining K. Then $\mathcal{O}^*$ consists of $n + 2 = 2m + 2$ distinct points, no three collinear. The lines of π form two classes with respect to $\mathcal{O}^*$, since each line of π is either a secant of $\mathcal{O}^*$ or a non-secant of $\mathcal{O}^*$. Each point of π which is not on $\mathcal{O}^*$ lies on exactly $m + 1$ secants of $\mathcal{O}^*$ and m non-secants of $\mathcal{O}^*$. The complete oval $\mathcal{O}^*$, together with its secants, is the design D. The non-secant lines of $\mathcal{O}^*$, together with the points of π not on $\mathcal{O}^*$, form a complementary design D'.

It should now be easy to check that, for D with parameters (2. 2), the existence of a complement is equivalent to the existence of a projective plane of order $n = 2m$ which possesses an oval (and thus a complete oval.)

The question as to whether every finite projective plane (of odd or even order) has an oval seems to have remained unanswered. There are non-isomorphic projective planes of the same order each having ovals, and there are many examples of a projective plane having differently embedded ovals. Thus, in the present context, D might have two non-isomorphic complements, but we could not decide without further examination whether the corresponding projective planes were isomorphic or not.

Our present intent is to start from D and try to construct a complement, D'. First, however, it is useful to begin with some complete ovals occurring in a projective plane over a field $F = GF(n)$ where

(2. 5) $\qquad\qquad n = 2m = 2^e , \ e > 1 .$

We choose integers s,t of form

(2. 6) $\qquad\qquad s = 2^f, \ t = 2^{e-f}, \ 1 \leq f < e .$

This ensures that the mappings

(2. 7) $\qquad\qquad x \to x^s , \ x \to x^t$

are automorphisms of $F = GF(n)$ distinct from the identity, and that each is the inverse of the other. We also want the mappings

(2. 8) $\qquad\qquad x \to x^{s-1} , \ x \to x^{t-1}$

to be one-to-one mappings of F upon F. The conditions for this to be true are

$$(n - 1, s - 1) = 1 = (n - 1, t - 1)$$

or, in view of (2. 7),

(2. 9) $\qquad\qquad\qquad (e, f) = 1 .$

Finally, to avoid duplications, we impose the equivalent conditions

(2. 10) $\qquad\qquad s \geq t ; \ e \leq 2f \leq 2e - 2 .$

R. H. Bruck

Now let us work in the affine plane A over F = GF(n) and con-
sider the affine curve $\mathscr{C}$(s) defined by either of the equivalent equations

(2, 11) $y = x^s$, $x = y^t$.

Here $\mathscr{C}$ (s) consists of the n distinct affine points (x, y) of form
(k, k^s) where k ranges over F. For any fixed k, c ϵ F with c $\neq$ 0, the
line

$$y = c(x - k) + k^s$$

meets $\mathscr{C}$(s) in the point (k, k^s) and in a second point $(k + u, k^s + u^s)$
where u ϵ F, u $\neq$ 0 is determined by

$$u^{k-1} = c .$$

The lines y = c, in the parallel class of the x-axis, and the lines x = c,
in the parallel class of the y-axis, each meet $\mathscr{C}$(s) in a unique point.
Hence, if $\mathscr{C}^*$(s) is the point-set of the corresponding projective plane
defined by

(2, 12) $\mathscr{C}^*$(s) = $\mathscr{C}$(s) $\cup$ {X,Y}   where

X = {y = c|c ϵ F} ,

Y = {x = c|c ϵ F} ,

are the "points at infinity" on the x-axis and y-axis respectively, then
$\mathscr{C}^*$(s) is a complete oval of the projective plane over GF(n).

It may be shown that, regardless of the (admissible) choice of s, t,
every collineation of the projective plane which maps $\mathscr{C}^*$(s) upon itself
must fix Y. Thus it is natural to think of

R. H. Bruck

$$\mathcal{C}(s) \cup \{X\}$$

as the oval and of Y as the knot. The group of collineations which fix X and $\mathcal{C}(s)$ is easily specified. There is a sharply doubly-transitive sub-group (on the points of $\mathcal{C}(s)$) consisting of all the mappings

(2.13) $\quad (x,y) \to (cx+d, c^s y+d^s)$, $c,d \in F$, $c \neq 0$;

and the full group is generated by the group (2.13) and the mapping

(2.14) $\qquad\qquad (x,y) \to (x^2, y^2)$.

When $t = 2$ (and not otherwise) there is also a collineation which moves X, namely

(2.15) $\qquad$ (For $t = 2$): $(0,0) \to X$, $X \to (0,0)$,

$$Y \to Y,$$

$$(x,y) \to (x^{-1}, x^{-1}y) \text{ for } x \neq 0.$$

Consequently, for $t = 2$, the group generated by the group (2.13) together with the mappings (2.15) is sharply triply transitive on $\mathcal{C}(s) \cup \{X\}$.

For $e \leq 4$, the above conditions require $t = 2$. However, for $e = 5$ we can have both

$$s = 16, \ t = 2$$

and

$$s = 8, \ t = 4.$$

Thus, for $e \geq 5$, we can expect differently imbedded complete ovals.

There are also the ovals of Tits, Luneburg and Suzuki, but we will not go into them here. See Luneburg [10].

3. (k = 2 continued.) Construction of complements. As in section 2, D is a design with parameters

(3.1) $v = 2m + 2$, $b = (m + 1)(2m + 1)$, $k = 2$, $r = 2m + 1$, $\lambda = 1$.

If there exists a complement, D', of D, then D' has parameters

(3.2) $v' = m(2m - 1)$, $b' = 4m^2 - 1$, $k' = m$, $r' = 2m + 1$, $\lambda' = 1$,

and the projective plane defined by D, D' has order

(3.2a) $n = 2m$.

Here m is a positive integer and, to avoid trivialities, we assume m > 1.

We may assume that the varieties, or points, of D are the first $v = 2m + 2$ positive integers. The blocks (or lines), the spreads and packings of D may be expressed in terms of elements of the symmetric group on these points as follows: A line is a transposition. A spread is an involution without fixed points; that is, a product of $v/2 = m + 1$ disjoint transpositions. A packing is a set of $r = 2m + 1$ distinct involutions without fixed points, such that the product of any two distinct involutions of the set also has no fixed points.

The numbers, s(m), of distinct spreads of D can be written down at once:

(3.3) $$s(m) = \frac{(2m + 2)!}{2^{m+1}(m + 1)!} = \prod_{i=0}^{m} (2i + 1) = (2m + 1)!! \ .$$

We notice that, for $m \geq 2$,

(3.4) $s(m)/b' = (2m - 3)!!$, $m \geq 2$.

Since, in constructing D', we use just b' (suitably chosen) spreads, we

R. H. Bruck

see that for m large we need only a small fraction of the spreads. But for m = 2, we need all of them.

It might be interesting to have a formula for the number, p(m), of distinct packings of D. As we shall see,

(3.5) $\qquad p(1) = 1 , \ p(2) = 6 , \ p(3) \geq 2040 .$

Now we shall consider the details for small values of m, beginning with the excluded case m = 1.

I. $\underline{m = 1}$. Here

(3.6)
$$v = 4 , \ b = 6 , \ k = 2 , \ r = 3 , \ \lambda = 1 ,$$
$$v' = 1 , \ b' = 3 , \ k' = 2 , \ r' = 3 , \ \lambda' = 1 ,$$
$$m = 1 , \ n = 2 .$$

We think of the four points 1, 2, 3, 4 as vertices of a planar quadrangle. There are 3 spreads, namely

$$(12)(34) , \ (13)(24) , \ (14)(23) .$$

These represent the diagonal points of the quadrangle. They also form the unique packing. The latter represents the line joining the diagonal points in a projective plane of order 2.

II. $\underline{m = 2}$. Here

(3.7)
$$v = v' = 6 , \ b = b' = 15 , \ k = k' = 2 , \ r = r' = 15 , \ \lambda = \lambda' = 1 ,$$
$$m = 2 , \ n = 4 .$$

And, as pointed out, the total number of spreads is $s(2) = 5!! = 15$.

To find the spreads and packings, we may divide the six points

R. H. Bruck

$1, 2, 3, 4, 5, 6$ into two unordered triangles, say $\{1, 2, 3\}$ and $\{4, 5, 6\}$.
If a spread contains a line of one triangle, then it contains one line of
each triangle and a single "cross-join", a line joining the remaining points
of the two triangles. There are 9 such spreads, a typical one being

$$\theta = (12)(45)(36) .$$

Note that θ is uniquely determined by its single cross-join, (36). The
remaining 6 spreads each consist of 3 distinct cross-joins; and there are
2 such spreads for each cross-join. Thus there are just 4 spreads of this
second type disjoint from θ. Arranged in terms of the cross-join containing the point 3, these are:

$$\phi_1 = (15)(26)(34) , \qquad \psi_1 = (16)(25)(34) ,$$

$$\phi_2 = (16)(24)(35) , \qquad \psi_2 = (14)(26)(35) .$$

We note that ϕ_1, ϕ_2 are disjoint and ψ_1, ψ_2 are disjoint but, for
$i, j = 1, 2$, ϕ_i and ψ_j have a common line.

Now we consider a packing. This must consist of 5 disjoint
spreads which partition the lines. The spreads containing the lines (12),
(13), (23) of the first triangle must be of the first type - single cross-join -
and must be three disjoint spreads. The remaining two spreads cannot
contain a line of triangle $\{1, 2, 3\}$, and hence must be of the second type.
In particular, a packing containing θ must either contain ϕ_1, ϕ_2 or contain ψ_1, ψ_2, together with two more spreads of the first type. There are in
fact exactly two packings containing θ, namely

$$\{ \theta, \phi_1, \phi_2, (13)(46)(25) , (23)(56)(14) \}$$

and

R. H. Bruck

$$\{\,\theta, \psi_1, \psi_2,\ (13)(56)(24),\ (23)(46)(15)\}\ ,$$

and these have only θ in common. If we note, in addition, that the four newly displayed spreads are the only ones of the first type disjoint from θ, we will have that to each spread α disjoint from θ there corresponds one and only one packing containing θ and α. Since every spread is a conjugate of θ in the symmetric group on $1, 2, 3, 4, 5, 6$, we have completed our investigation.

We may use the above work to show the uniqueness of the projective plane of order 4. First we verify, by a counting argument, that such a plane must contain a set of six distinct points, no three collinear. Then, by the above analysis, the plane is uniquely determined by the six points.

III. $\underline{m = 3.}$ Here

$$(3.8) \quad \begin{aligned} & v = 8,\ b = 20,\ k = 2,\ r = 7,\ \lambda = 1\,, \\ & v' = 15,\ b' = 63,\ k' = 3,\ r' = 7,\ \lambda' = 1\,, \\ & m = 3,\ n = 6\,. \end{aligned}$$

Also $s(3) = 105.$

We know from the beginning that D has no complement D', since there is no projective plane of order 6. Nevertheless the spreads and packings of D are worth studying, both in themselves and for use with larger values of m.

First we shall consider a type of packing which may be called projective. Behind this concept is the fact that, in many ways, the 8 points of D can be put into one-to-one correspondence with the 8 points of affine 3-space over GF(2). Once such a correspondence has been chosen, the lines of D fall into 7 disjoint parallel classes of 4 lines each. Each such parallel class is uniquely specified by a spread of D. The 7 spreads form a packing which represents the plane at infinity, with the 7 spreads as its points. Finally, the 7 spreads fall into 7 distinct partial packings of 3 spreads each which represent the 7 lines at infinity.

Let us remark before going on that a similar possibility arises in case $k = 2$, $v = 2^e$, $e > 2$. In this case a projective packing would represent, not a plane, but the hyperplane at infinity of an affine e-space over $GF(2)$. It would uniquely determine projective e-space over $GF(2)$, since it would specify all the parallel classes of lines of affine e-space.

An intrinsic definition, adequate for the present case $(v = 8)$, may be given as follows: A packing, $\mathcal{P}$, of D is called _projective_ provided that, for any four distinct points a, b, c, d of D, if there is a spread θ in $\mathcal{P}$ containing the lines (ab), (cd), then there is a spread ϕ in $\mathcal{P}$ containing the lines (ac), (bd). (Thus also, on interchange of c, d, there is a spread ψ in $\mathcal{P}$ containing the lines (ad), (bc).) Reason for the definition: If $\mathcal{P}$ represents the plane at infinity of affine 3-space over $GF(2)$, then θ represents a parallel class of lines and hence a, b, c, d are the four points of an affine plane (of order 2.)

We note the following:

(i) _Each spread, θ, of D is contained in exactly two distinct projective packings. The two packings have only θ in common._

To see this, consider the spread

$$\theta = (12)(34)(56)(78) .$$

Applying the criterion to the four points $1, 2, 3, 4$ and also to the four points $5, 6, 7, 8$, we see that a projective packing containing θ must contain spreads of form

R. H. Bruck

$$\phi = (13)(24)\alpha \, , \qquad \psi = (14)(23)\beta$$

where α, β are (57)(68), (58)(67) in one of the two orders. We shall go on with the case that

$$\theta = (12)(34)(56)(78) \, ,$$
$$\phi = (13)(24)(57)(68) \, ,$$
$$\psi = (14)(23)(58)(67) \, .$$

In any packing there must be a spread λ containing (15). If the spread is projective and contains θ, ϕ, ψ, then, by reference to θ, ϕ, ψ in turn, λ must contain (26), (37), (48). Hence

$$\lambda = (15)(26)(37)(48) \, .$$

Continuing in this way we arrive at the following packing:

$\mathcal{P}$: (12)(34)(56)(78) = θ
  (13)(24)(57)(68)
  (14)(23)(58)(67)
  (15)(26)(37)(48)
  (16)(25)(38)(47)
  (17)(28)(35)(46)
  (18)(27)(36)(45) .

We need to verify that $\mathcal{P}$ is projective, but this is indeed the case, and $\mathcal{P}$ is the unique projective packing containing θ, ϕ, ψ. The second projective packing, $\mathcal{P}'$, containing θ may be obtained in similar fashion or, alternatively, by transforming $\mathcal{P}$ by the permutation (78). Clearly $\mathcal{P}$ and $\mathcal{P}'$ have only θ in common. This proves (i). Next:

R. H. Bruck

(ii) There are exactly 30 distinct projective packings of D. These form a single orbit under the symmetric group S_8. If $\mathcal{P}$ is a projective packing and $G(\mathcal{P})$ is the subgroup of S_8 mapping $\mathcal{P}$ upon $\mathcal{P}$, then $G(\mathcal{P})$ is sharply transitive on ordered tetrahedra. In particular,

$$|G(\mathcal{P})| = 8 \cdot 7 \cdot 6 \cdot 4 .$$

To begin with, we know that each of the 105 spreads is contained in exactly two distinct projective packings. On the other hand, each packing contains exactly 7 spreads. Hence there are exactly

$$105. \, 2/7 = 30$$

distinct projective spreads.

Let θ be a spread and let $\mathcal{P}$, $\mathcal{P}'$ be the two distinct projective packings containing θ. Let ϕ be a spread contained in a projective packing $\mathcal{Q}$. There exists a permutation λ in S_8 such that

$$\lambda^{-1} \phi \lambda = \theta ;$$

consequently,

$$\lambda^{-1} Q \lambda = \mathcal{P} \text{ or } \mathcal{P}' .$$

If $\lambda^{-1} Q \lambda = \mathcal{P}'$ then, by the proof of (i), there exists a 2-cycle τ in S_8 such that

$$\tau^{-1} \lambda^{-1} Q \lambda \tau = \mathcal{P} .$$

Hence $\mathcal{Q}$ is a conjugate of $\mathcal{P}$. Thus the 30 projective packings form a conjugate class - a single orbit under S_8.

If $\mathcal{P}$ is a projective packing, there are several ways of determin-

R. H. Bruck

ing $G(\mathcal{P})$. One is as follows: $\mathcal{P}$ makes D into affine 3-space over GF(2), and $G(\mathcal{P})$ is the corresponding group of collineations. We can pick an ordered triple a, b, c of distinct points of D in $8 \cdot 7 \cdot 6$ distinct ways. The plane of a, b, c (with respect to $\mathcal{P}$) contains a unique fourth point. Hence we can pick a point d, not in this plane, in 4 ways. That is, the number of ordered tetrahedra

$$a, b, c, d$$

of D is $8 \cdot 7 \cdot 6 \cdot 4$. Given such a tetrahedron, the remaining four points of D lie one each on the four faces of the tetrahedron. Hence, clearly, $G(\mathcal{P})$ is sharply transitive on ordered tetrahedra and has the stated order. This proves (ii).

We could also calculate $|G(\mathcal{P})|$ as follows: Since the 30 projective spreads form a single orbit under S_8 then

$$30 = |S_8 : G(\mathcal{P})|$$

and hence

$$|G(\mathcal{P})| = 8!/30 = 8 \cdot 7 \cdot 4! = 8 \cdot 7 \cdot 6 \cdot 4 .$$

Before we introduce the next type of packing, we shall investigate the orbits of the spreads of D under conjugation by the cyclic group $\langle \pi \rangle$ generated by

$$\pi = (2\,3\,4\,5\,6\,7\,8) .$$

The 105 spreads break up into 15 orbits of length 7. A straightforward calculation shows that exactly three of these orbits are packings, namely:

R. H. Bruck

$\mathcal{P}_1$: orbit of (12)(35)(48)(67) under $\langle \pi \rangle$;

$\mathcal{P}_2$: orbit of (12)(37)(45)(68) under $\langle \pi \rangle$;

$\mathcal{Q}$: orbit of (12)(38)(47)(56) under $\langle \pi \rangle$.

The packings $\mathcal{P}_1$, $\mathcal{P}_2$ are easily verified to be projective. The remaining packing,

$\mathcal{Q}$: (12)(38)(47)(56)

(13)(24)(58)(67)

(14)(26)(35)(78)

(15)(28)(37)(46)

(16)(23)(48)(57)

(17)(25)(34)(68)

(18)(27)(36)(45)

is not projective.

Let σ be the unique element of S_8 which fixes $1, 2$ and satisfies

$$\sigma^{-1} \pi \sigma = \sigma^3 .$$

Thus

$$\sigma = (3\,5\,4\,8\,6\,7) .$$

Clearly σ must permute the packings $\mathcal{P}_1$, $\mathcal{P}_2$, $\mathcal{Q}$. In fact

$$\sigma^{-1} \mathcal{P}_1 \sigma = \mathcal{P}_2 , \quad \sigma^{-1} \mathcal{P}_2 \sigma = \mathcal{P}_1$$

and

$$\sigma^{-1} \mathcal{Q} \sigma = \mathcal{Q} .$$

R. H. Bruck

Furthermore, σ fixes the first spread of $\mathcal{Q}$ and permutes the remaining six spreads of $\mathcal{Q}$ in a single cycle.

It is easy to see that the only permutation in $G(\mathcal{Q})$ which fixes 1, 2 and 3 is the identity. Again, if $G(\mathcal{Q})$ contains an element which moves 1, then, in view of the properties of π, σ, $G(\mathcal{Q})$ must be sharply triply transitive on D. In particular, $G(\mathcal{Q})$ must contain a cycle of length six which fixes 3, 4 and maps 1 upon 2. A straightforward search ends in a contradiction. Hence:

(iii) $G(\mathcal{Q})$ is a non-abelian group of order 42 such that

(iii)(a) $G(\mathcal{Q})$ fixes 1 and is sharply doubly-transitive on the remaining 7 elements of D and

(iii)(b) $G(\mathcal{Q})$ is sharply doubly-transitive on the 7 spreads of $\mathcal{Q}$.

We shall call any packing equivalent to $\mathcal{Q}$ a non-projective cyclic packing.

(iv) Let $\mathcal{R}$ be a packing of D such that $G(\mathcal{R})$ is transitive on the 7 spreads of $\mathcal{R}$. Then $\mathcal{R}$ is either a projective packing or a non-projective cyclic packing.

Proof. Since $G = G(\mathcal{R})$ is transitive on the spreads of $\mathcal{R}$,

$$|G(\mathcal{R})| = 7n$$

where n is the order of the subgroup of G fixing a spread of $\mathcal{R}$. Since the centralizer in S_8 of a spread has order $2^4 \cdot 4!$,

$$n \mid 2^7 \cdot 3 .$$

A Sylow 7-group of G has order 7 and is a subgroup of S_8. Hence, after replacing $\mathcal{R}$ by a conjugate, we may assume that G has a Sylow 7-group generated by the permutation π (see above). In this case, $\mathcal{R}$ must be $\mathcal{P}_1$, $\mathcal{P}_2$ or $\mathcal{Q}$. This proves (iv). We may verify further that

$$n = 2^6 \cdot 3 \text{ or } 2 \cdot 3 .$$

(v) <u>There are exactly 960 distinct non-projective cyclic packings</u> <u>of D, and these form a single orbit under</u> S_8.

<u>Proof.</u> It is enough to show that S_8 has exactly 960 distinct Sylow 7-subgroups. Such a subgroup fixes a unique point a of D; and a can be chosen in 8 ways. The number of 7-cycles fixing a is 6!, and these lie in sets of 6 in the Sylow 7-subgroups fixing a. Hence there are precisely

$$8 \cdot 5! = 960$$

Sylow 7-subgroups. This proves (v).

To make further progress, we need the following:

(vi) <u>If θ is a spread of D, there are precisely 60 distinct</u> <u>spreads of D disjoint from θ. These split into two orbits, of lengths 12</u> <u>and 48, under conjugation by the centralizer of θ in S_8. The orbit of</u> <u>length 12 consists of all spreads of D which are disjoint from θ and</u> <u>commute with θ; this orbit also consists of</u> $\mathcal{P} \cup \mathcal{P}\prime - \{\theta\}$ <u>where</u> $\mathcal{P}, \mathcal{P}\prime$ <u>are the two regular packings of D which contain θ. If λ is in</u> <u>the orbit of length 48, then $\theta\lambda$ is the product of two disjoint cycles of</u> <u>length 4, and $(\theta\lambda)^2$ is the unique spread of D which is disjoint from and</u> <u>commutes with each of θ, λ.</u>

<u>Proof.</u> To begin with, for $0 \leq i \leq 4$, let $f(i)$ be the number of

R. H. Bruck

spreads of D which are distinct from θ and have exactly i lines in common with θ. It should be obvious that

$$f(3) = f(4) = 0 ,$$
$$f(0) + f(1) + f(2) = 104 .$$

Next we calculate f(2). We can choose two distinct lines of θ in 6 ways. Once these are chosen, we can construct a spread having precisely these two lines in common with θ in 2 ways. Hence

$$f(2) = 12 ,$$
$$f(0) + f(1) = 92 .$$

Next we calculate f(1). We can choose a line of θ in 4 ways. The number of spreads of D having only this line in common with θ is

$$5 \cdot 3 - 1 - 3 \cdot 2 = 8 .$$

Hence

$$f(1) = 32, \ f(0) = 60 .$$

Thus there are precisely $f(0) = 60$ spreads of D disjoint from θ.

Next let ϕ be a spread disjoint from θ which commutes with θ. Then, under conjugation, ϕ must permute the 4 lines of θ in two cycles of length 2. Equivalently, ϕ must be one of the 12 spreads in $\mathcal{P} \cup \mathcal{P}' - \{\theta\}$. From previous work, or by direct calculation, the latter set forms a single orbit under the centralizer of θ in S_8.

For the last part of the proof, we may assume that

$$\theta = (12)(34)(56)(78) .$$

The spread

$$\lambda_0 = (13)(25)(47)(68)$$

is disjoint from θ. Also

$$\theta\lambda_0 = (1584)(2376) \ ,$$

$$(\theta\lambda_0)^2 = (18)(27)(36)(45) \ .$$

Clearly $(\theta\lambda_0)^2$ is a spread which is disjoint from θ, λ_0 and commutes with both of them. The common centralizer of θ, λ_0 is easily seen to be sharply transitive on the elements of D. Thus it has order 8. Indeed, it consists of the identity permutation, the 5 spreads $(\theta\lambda_0)^2$ and

$$(\underline{12})(35)(46)(78) \ , \ (13)(24)(57)(\underline{68})$$

$$(16)(\underline{25})(38)(47) \ , \ (17)(28)(34)(\underline{56}) \ ,$$

and the elements

$$(1485)(2376) \ , \ \ (1584)(2673)$$

of order 4. Clearly the only spread in this common centralizer disjoint from θ, λ_0 is $(\theta\lambda_0)^2.$

The centralizer of θ has order

$$2^4 \cdot 4!$$

Hence the number of conjugates of λ_0 under this centralizer is

$$2^4 \cdot 4!/8 = 2 \cdot 4! = 48 \ .$$

This is enough for the proof of (vi).

R. H. Bruck

In addition we need:

(vii) <u>Let $\mathcal{P}, \mathcal{P}'$ be the two distinct projective packings contain-</u>
<u>ing a spread θ of D. Then each spread α in $\mathcal{P} - \{\theta\}$ is disjoint</u>
<u>from exactly 4 spreads in $\mathcal{P}' - \{\theta\}$, but commutes with none of these.</u>
<u>If β is any one of the 4 spreads in $\mathcal{P}' - \{\theta\}$ disjoint from α, then</u>
<u>$(\alpha\beta)^2 = \theta$.</u>

<u>Proof.</u> The first two sentences of (vii) require a straightforward
verification which we will omit. The last sentence then follows from (vi).

Now we come to:

(viii) <u>Let S be a commutative set of two or more disjoint spreads</u>
<u>of D. Then there exists one and only one projective packing, [S], of D,</u>
<u>which contains S.</u>

<u>Proof.</u> If α, β are two distinct (hence disjoint) elements of S,
then, since $\alpha\beta = \beta\alpha$, there exists a unique projective packing containing
α, β. Call this $[\alpha, \beta]$.

If $|S| = 2$, there is nothing more to prove.

If S has third element γ distinct (and disjoint) from α, β, then
there exists a unique projective packing $[\alpha, \gamma]$ containing α, γ. If
$[\alpha, \beta] \neq [\alpha, \gamma]$, then $[\alpha, \beta]$, $[\alpha, \gamma]$ are the two distinct projective packings containing α. Hence, by (vii), $\beta\gamma \neq \gamma\beta$, a contradiction. Therefore
(and similarly)

$$[\alpha, \beta] = [\alpha, \gamma] = [\beta, \gamma].$$

If $|S| = 3$, we are done. If S has a fourth element δ, distinct
from α, β, γ, then

R. H. Bruck

$$[\alpha, \beta] = [\beta, \gamma] = [\gamma, \delta] .$$

This shows that every two distinct elements of S are contained in the same (unique) projective packing, which we may call [S]. Thus (viii) is true.

(ix) Let S be a commutative set of 6 disjoint spreads of D. Then there is one and only one packing of D containing S, namely the projective packing [S].

Proof. Let θ be the unique element of [S] which is not in S. If a packing contains S and a seventh spread ϕ, then θ, ϕ cover the same lines of D, so $\phi = \theta$. This proves (ix).

(x) Let S be a commutative set of 5 disjoint spreads of D. Then there is exactly one packing of D which contains S but is not projective. The set of all packings of D which contain a commutative set of 5 disjoint spreads but are not projective consists of precisely 630 distinct spreads and forms a single orbit under conjugation by S_8.

Proof. Let θ, ϕ be the two distinct spreads which are in [S] but not in S. In view of (vi), after replacing S by a suitable conjugate, we may assume without loss of generality that

$$\theta = (12)(34)(56)(78) ,$$
$$\phi = (13)(24)(57)(68) .$$

We wish to form a new packing by replacing θ, ϕ in [S] by two new spreads θ', ϕ' which cover the same lines as θ, ϕ. We may suppose that θ', ϕ' contain (12), (13) respectively. Then ϕ' cannot contain (34) and θ' cannot contain (24). Therefore

R. H. Bruck

$$\theta' = (12)(34)\lambda \ ,$$
$$\phi' = (13)(24)\mu \ ,$$

where λ, μ are disjoint and each contain two disjoint lines chosen from

$$(56), (78), (57), (68) \ .$$

Clearly λ, μ must be

$$(56)(78), \quad (57)(68)$$

in some order. If $\lambda = (56)(78)$, we get $\theta' = \theta$, $\phi' = \phi$, contrary to desire. Hence we must have

$$\theta' = (12)(34)(57)(68) \ ,$$
$$\phi' = (13)(24)(56)(78) \ ,$$

and

$$\theta'\phi' = \phi'\theta' = \theta\phi = \phi\theta = (14)(23)(58)(67) \ .$$

The packing obtained by replacing θ, ϕ in [S] by θ', ϕ' is not projective, since it does not contain θ, ϕ, whereas the projective packing [S] is determined by any two distinct spreads in S. Note, also, that of the two projective packings of D containing the element $\theta\phi$ of S, one is [S] and the other contains θ', ϕ'. Hence each of θ', ϕ' commutes with exactly one element of S, namely $\theta\phi$.

Conversely, let θ, ϕ be the spreads displayed in the preceding paragraph, so that $[\theta, \phi]$ is the projective packing $\mathcal{P}$ displayed in the proof of (i). Then S is the set of spreads of $\mathcal{P}$ distinct from θ, ϕ, and we are interested in the packing $\mathcal{R}$ obtained from $\mathcal{P}$ by replacing θ, ϕ by θ', ϕ'. Clearly $G(\mathcal{R})$ is isomorphic to that subgroup of $G(\mathcal{P})$

which fixes the unordered pair θ, ϕ. Hence, by (ii), $G(\mathcal{R})$ is sharply transitive on those ordered tetrahedra

$$a, b, c, d$$

of D (with respect to $\mathcal{P}$) such that one of the lines (ab), (ac) is in θ and the other is in ϕ. Therefore

$$|G(\mathcal{R})| = 8 \cdot 2 \cdot 1 \cdot 4 = 8 \cdot 4 \cdot 2$$

and the number of conjugates of $\mathcal{R}$ under S_8 is

$$|S_8 : G(\mathcal{R})| = 7 \cdot 6 \cdot 5 \cdot 3 = 630 .$$

This completes the proof of (x).

Next we need two simple concepts. Let $\{\alpha, \beta, \gamma\}$ be a commutative set of three disjoint spreads and let

$$\mathcal{P} = [\alpha, \beta, \gamma]$$

be the unique projective packing containing α, β, γ. Then $\mathcal{P}$ is the (projective) plane at infinity, and the spreads of $\mathcal{P}$ are the points at infinity, of the affine 3-space of D determined by $\mathcal{P}$. It may be verified that the line at infinity containing α, β consists of the three disjoint spreads α, β and $\alpha\beta = \beta\alpha$. Hence we make the following definitions: $\{\alpha, \beta, \gamma\}$ is a line at infinity if $\alpha\beta = \gamma$ and is a triangle at infinity if $\alpha\beta \neq \gamma$.

(xi) There are precisely 420 distinct packings $\mathcal{R}$ of D such that $\mathcal{R}$ contains a maximal commutative subset S of 4 disjoint spreads of D. These packings form a single orbit under conjugation by S_8. Thus, if $\mathcal{R}$ is such a packing,

R. H. Bruck

$$|G(\mathcal{R})| = 8 \cdot 4 \cdot 3 .$$

Moreover, if S is a commutative set of 4 disjoint spreads of D, a necessary and sufficient condition that S be imbeddable as a maximal commutative subset of a packing $\mathcal{R}$ is that the complement of S in the projective packing [S] be a line at infinity. When the condition is satisfied, $\mathcal{R}$ can be chosen in two distinct ways; the two packings have only S in common.

Proof. Let S be a commutative set of 4 disjoint packings. We can assume without loss of generality that [S] = $\mathcal{P}$ where $\mathcal{P}$ is the projective packing exhibited in the proof of (i). Let K be the complement of S in $\mathcal{P}$.

Case 1. K is a triangle at infinity. Then, by (ii), we can assume without loss of generality that K consists of the disjoint spreads

$$\theta = (12)(34)(56)(78) ,$$
$$\phi = (13)(24)(57)(68) ,$$
$$\chi = (15)(26)(37)(48) .$$

If S can be imbedded as a maximal commutative subset in a packing $\mathcal{R}$, then $\mathcal{R}$ must be obtained from $\mathcal{P}$ by replacing θ, ϕ, χ by three disjoint spreads

$$\theta' = (12) \ldots ,$$
$$\phi' = (13) \ldots ,$$
$$\chi' = (15) \ldots ,$$

which cover the same lines as θ, ϕ, χ but are distinct from θ, ϕ, χ. Clearly one of the lines (24), (26) must be in ϕ' and the other in χ.

R. H. Bruck

There are two cases:

Case 1.1. $\theta = (12)\ldots$,

$\phi' = (13)(24)\ldots$,

$\chi' = (15)(26)\ldots$.

Clearly the line (57) must not be in ϕ', (else we will have $\phi' = \phi$) and cannot be in χ'. Therefore (57) must be in θ: Similarly, (37) cannot be in χ', else $\chi' = \chi$, and cannot be in ϕ'. This puts both (57), (37) in θ', a contradiction. Hence Case 1.1 cannot occur.

Case 1.2. $\theta' = (12)\ldots$,

$\phi' = (13)(26)\ldots$,

$\chi' = (15)(24)\ldots$.

The line (34) cannot be in ϕ' or χ', hence must be in θ'. Then (37) cannot be in θ' or ϕ', hence must be in χ'. This gives

$\theta' = (12)(34)\ldots$,

$\phi' = (13)(26)\ldots$,

$\chi' = (15)(24)(37)(68)$.

Thus (48) must be in ϕ' and hence (78) must be in θ'. But then

$\theta' = (12)(34)(56)(78) = \theta$,

a contradiction. Hence Case 1.2 and Case 1 are impossible. That is: K is not a triangle at infinity. This leaves:

Case 2. K is a line at infinity. Then we may assume without loss of generality that K consists of

R. H. Bruck

$$\theta = (12)(34)(56)(78) ,$$

$$\phi = (13)(24)(57)(68) ,$$

$$\psi = (14)(23)(58)(67) .$$

We see easily that three disjoint spreads covering the same lines as θ, ϕ, ψ must be made up by pairing the three products

$$(12)(34), \ (13)(24), \ (14)(23)$$

with the three products

$$(56)(78), \ (57)(68), \ (58)(67)$$

in some order. We may regard θ, ϕ, ψ as giving the normal pairing. Then, if the new spreads are to be distinct from θ, ϕ, ψ we see that these are just two solutions, corresponding to the two derangements of the normal pairing: Either we can replace θ, ϕ, ψ in $\mathcal{P}$ by

$$\theta' = (12)(34)(57)(68) ,$$

$$\phi' = (13)(24)(58)(67) ,$$

$$\psi' = (14)(23)(56)(78) ,$$

getting a packing $\mathcal{R}_1$, or we can replace θ, ϕ, ψ in $\mathcal{P}$ by

$$\theta'' = (12)(34)(58)(67) ,$$

$$\phi'' = (13)(24)(56)(78) ,$$

$$\psi'' = (14)(23)(57)(68) ,$$

getting a packing $\mathcal{R}_2$.

In each of $\mathcal{R}_1, \mathcal{R}_2, S$ occurs as a maximal commutative subset.

Let H be the normalizer in S_8 of the subset

R. H. Bruck

$$K = \{\theta, \phi, \psi\} \, .$$

Since $[K] = \mathcal{P}$, H is a subgroup of $G(\mathcal{P})$. Hence, by (ii), it is clear that H is sharply transitive on the ordered tetrahedra

$$a, b, c, d$$

of D (relative to $\mathcal{P}$) such that the lines (ab), (ac) belong to some two of θ, ϕ, ψ. Thus

$$|H| = 8 \cdot 3 \cdot 2 \cdot 4 = 8 \cdot 4 \cdot 3 \cdot 2 \, .$$

Clearly each element of H either normalizes both of the sets

$$\{\theta', \phi', \psi'\} \, , \quad \{\theta'', \phi'', \psi''\}$$

or interchange the two sets. The permutation

$$(15)(26)(37)(48)$$

is in H and interchanges the two sets. Consequently $\mathcal{R}_1$, $\mathcal{R}_2$ are conjugate under H (and S_8!) and

$$G(\mathcal{R}_1) = G(\mathcal{R}_2)$$

is a subgroup of index 2 in H. Thus

$$|G(\mathcal{R}_1)| = 8 \cdot 4 \cdot 3$$

and

$$|S_8 : G(\mathcal{R}_1)| = 7 \cdot 6 \cdot 5 \cdot 2 = 420 \, .$$

This completes the proof of (xi).

R. H. Bruck

The following statement throws additional light on the packings described in (x) and (xi).

(xii) _Let A be a commutative set of 4 distinct spreads of D such that the complement, L, of A in the projective packing [A] is a line at infinity. Then a line at infinity, T, can be chosen in precisely 5 distinct ways such that T is disjoint from A and the set_ $\mathcal{R}$ = T ∪ A _is a packing which contains T as a maximal commutative subset. In two cases, T is disjoint from L and A is a maximal commutative subset of_ $\mathcal{R}$. (Compare (xi).) _In the remaining three cases, T meets L in a spread, say_ α, _and_ S = A ∪ {α} _is a maximal commutative subset of_ $\mathcal{R}$. (Compare (x).)

Sketch of proof. If T is to be disjoint from L, we consider the proof of (xi) with S = A.

If T is to meet L in a spread α (which can be chosen in 3 ways) we consider the proof of (x) with S = A ∪ {α}.

The following gives still another look at (x):

(xiii) _Let_ α,β _be disjoint spreads of D such that_ αβ ≠ βα. _Let L, M be the unique intersecting lines at infinity which contain_ α,β _respectively. Thus_

$$L = \{\alpha, \beta\alpha\beta, (\alpha\beta)^2\}, \quad M = \{\beta, \alpha\beta\alpha, (\alpha\beta)^2\}.$$

Then there are exactly two distinct packings of D which contain L ∪ M. In one of these, L is a maximal commutative subset and M is contained in a maximal commutative subset of 5 distinct spreads. In the other the roles of L, M are interchanged.

Proof. We may assume without loss of generality that

R. H. Bruck

$$\alpha = (12)(34)(56)(78) ,$$
$$\beta = (13)(25)(47)(68) .$$

Then we find that

$$\alpha\beta\alpha = (16)(24)(38)(57) ,$$
$$\beta\alpha\beta = (17)(28)(35)(46) ,$$
$$(\alpha\beta)^2 = (18)(27)(36)(45) .$$

Thus the 5 spreads in $L \cup M$ are disjoint but do not cover the lines

$$(14) ,(15); (48),(58); (23),(26); (37),(67) .$$

Hence the remaining disjoint spreads γ, δ of a packing containing $L \cup M$ must pair the products

$$(14)(58), (15)(48)$$

with the products

$$(23)(67), (26)(37) .$$

This gives the two packings

$$\mathcal{R}_1 = L \cup M \cup \{\gamma_1, \delta_1\} ,$$
$$\mathcal{R}_2 = L \cup M \cup \{\gamma_2, \delta_2\} ,$$

where

$$\gamma_1 = (14)(58)(23)(67) , \quad \gamma_2 = (14)(58)(26)(37) ,$$
$$\delta_1 = (15)(48)(26)(37) , \quad \delta_2 = (15)(48)(23)(67) .$$

It is easy to check that, in $\mathcal{R}_1$, M and $L \cup \{\gamma_1, \delta_1\}$ are maximal com-

R. H. Bruck

mutative subsets and, in $\mathcal{R}_2$, L and $M \cup \{Y_2, \delta_2\}$ are maximal commutative subsets. This proves (xiii).

Next we shall need the following:

(xiv) Let α, β be disjoint spreads of D such that $\alpha\beta \neq \beta\alpha$.
Then there are precisely 31 distinct spreads of D disjoint from both of
α, β. Of these:

(a) 1 (namely $(\alpha\beta)^2$) commutes with both of α, β;

(b) 5 commute with α but not with β;

(c) 5 commute with β but not with α;

(d) 20 commute with neither of α, β.

Sketch of proof. As in the proof of (vi), for $0 \leq i \leq 4$, let $f(i)$
be the number of spreads of D which are distinct from α and have
exactly i lines in common with α. In addition, for $0 \leq i, j \leq 4$, let
$g(i, j)$ be the number of spreads of D which are distinct from α, β and
have exactly i, j lines in common with α, β respectively. By the proof
of (vi), and by symmetry,

$$g(i, j) = 0 \text{ if } i \geq 3 \text{ or } j \geq 3 ;$$
$$g(i, j) = g(j, i), \text{ all } i, j ;$$

and

$$g(2, 2) + g(2, 1) + g(2, 0) = f(2) = 12 ,$$
$$g(1, 2) + g(1, 1) + g(1, 0) = f(1) = 32 ,$$
$$g(0, 2) + g(0, 1) + g(0, 0) = f(0) - 1 = 59 .$$

Next we specialize α, β (as warranted by (vi)) and verify that

R. H. Bruck

$$g(2,2) = 0 \; ; \; g(2,1) = g(1,2) = 4 \; ;$$
$$g(2,0) = g(0,2) = g(1,1) = 8 \; .$$

From these values in the above equations,

$$g(1,0) = g(0,1) = 20$$

and

$$g(0,0) = 31 \; .$$

This proves the first statement of (xiv). To prove (a), (b) we need merely re-examine the 12 spreads disjoint from α and commuting with α in relation to β. (Compare (vi), (vii).) And (c), (d) follow from (a), (b). This proves (xiv).

We shall call a set S of n disjoint spreads of D <u>anti-commuta-</u> <u>tive</u> if either n = 1 or n > 1 and $\alpha\beta \neq \beta\alpha$ for every two distinct spreads α, β in S.

(xv) <u>Let S be an anti-commutative set of n disjoint spreads. Then the number, N(n), of non-projective cyclic packings of D which contain S, is given as follows:</u>

$$N(1) = 64; \; N(2) = 8; \; N(3) = 1;$$
$$N(4) = N(6) = N(7) = 1 \; .$$

<u>However, N(5) = 1 or 0, depending on the case.</u>

Sketch of proof. We begin by noting that the non-projective cyclic packing $\mathcal{Q}$, displayed after the proof of (iv), is an anti-commutative set of 7 disjoint spreads. From this and (iv), (vi), we have

R. H. Bruck

$$105 \cdot N(1) = 960 \cdot 7 ,$$

$$105 \cdot 48 \cdot N(2) = 960 \cdot 7 \cdot 6 ,$$

whence $N(1) = 64$, $N(2) = 8$.

Next we note that the first two spreads of $\mathcal{Q}$ are the disjoint pair

$$\alpha = (12)(38)(47)(56) ,$$

$$\beta = (13)(24)(58)(67) .$$

We need, in effect, to examine the 8 cyclic packings containing α, β and, for this, the following considerations prove useful. Since $\alpha\beta \neq \beta\alpha$, it turns out that for each ordered pair d, d' of elements of D (not necessarily distinct) there is one and only one permutation of D which interchanges α, β (by conjugation) and maps d upon d'. In particular, the eight permutations interchanging α, β are the four odd powers of

$$\Delta = (1 2 4 7 6 5 8 3)$$

together with

$$\lambda = (23)(48)(57), \text{ fixing } 1, 6 ;$$

$$\mu = (14)(37)(68), \text{ fixing } 2, 5 ;$$

$$\nu = (16)(27)(35), \text{ fixing } 4, 8 ;$$

$$\rho = (18)(25)(46), \text{ fixing } 7, 3 .$$

Clearly, under conjugation, Δ induces the cycle

$$(\lambda, \mu, \nu, \rho) .$$

Each cyclic packing containing α, β must be the orbit of α under successive conjugation by a permutation of D which fixes one

R. H. Bruck

element of D, permutes the remaining elements of D in a cycle of length
7, and maps α upon β. However (as may be verified) there are, in all,
8 distinct cycles of length 7 which fix 1 and map α upon β, but for only
one is the orbit of α a packing. The correct choice is simply described;
namely, it is

$$(12)\alpha\lambda = (2\ 3\ 4\ 5\ 6\ 7\ 8) = \pi \ .$$

Here $(12)\alpha$ is obtained from α by dropping the line through 1, and λ
is the unique permutation which interchanges α, β and fixes 1. The other
cycles for which the orbit of α is a cyclic packing containing α, β can
be written down analogously. We prefer, however, to use the 8 permutations

$$\pi_i = \Delta^{-i}\pi\Delta^i, \quad 0 \leq i \leq 7 \ .$$

Each of these is a cycle of length 7, and no two fix the same point of D.
Also π_i maps α upon β or β upon α according as i is even or odd.
Finally, the orbit

$$\mathcal{Q}_i = \Delta^{-i}\mathcal{Q}\Delta^i, \quad 0 \leq i \leq 7 \ ,$$

of $\mathcal{Q}$ under conjugation by Δ^i, is a (non-projective) cyclic packing
containing α, β.

Next we define

$$\theta_k = \pi^{-k}\alpha\pi^k, \quad 0 \leq k \leq 6 \ ,$$

so that $\theta_0 = \alpha$, $\theta_1 = \beta$, and the θ_k make up $\mathcal{Q} = \mathcal{Q}_0$. We also define

R. H. Bruck

$$A_i = \Delta^{-i} \theta_2 \Delta^i , \quad B_i = \Delta^{-i} \theta_3 \Delta^i ,$$

$$C_i = \Delta^{-i} \theta_4 \Delta^i$$

for all integers i and verify that the elements

$$A_i, B_i \quad (0 \leq i \leq 7)$$

and

$$C_0, C_1, C_2, C_3$$

are the 20 distinct spreads disjoint from α, β which do not commute with α or β. In particular,

(*) $$A_8 = A_0, \ B_8 = B_0, \ C_4 = C_0 .$$

In addition we find that

$$A_1 = \theta_6, \ B_2 = \theta_5 .$$

Consequently we have:

$$\mathcal{Q}_0 = \mathcal{Q} : \ \alpha, \beta \ ; \ A_0, A_1 \ ; \ B_0, B_2 \ ; \ C_0 .$$

The other $\mathcal{Q}_i$ are obtained by successive addition of 1 to the subscripts of the A's, B's and C's, subject to (*).

Now we see that:

α, β, A_0 lie only in $\mathcal{Q}_0, \mathcal{Q}_7$;

α, β, B_0 lie only in $\mathcal{Q}_0, \mathcal{Q}_6$;

α, β, C_0 be only in $\mathcal{Q}_0, \mathcal{Q}_4$.

And this is enough to prove that $N(3) = 2$.

The rest of the proof can only be indicated. Of the 19 spreads disjoint from α, β, distinct from A_0, which commute with neither of α, β,

exactly 9 are disjoint from A_0, and one of these commutes with A_0. Of

the remaining 8, four lie on $\mathcal{Q}_0$ and four on $\mathcal{Q}_7$. Moreover, each of

the four on $\mathcal{Q}_0$ is disjoint from exactly one of the four on $\mathcal{Q}_7$.

The situation is similar for B_0 (or C_0), except that there are two

spreads (instead of one) disjoint from α, β and from B_0 (or C_0) which

do not commute with α or β but commute with B_0 (or C_0). •

Consequently, we may conclude that

$$N(4) = N(6) = N(7) = 1$$

and that $N(5) = 0$ or 1.

As an instructive example of an anti-commutative set of 5 disjoint

spreads, consider the following:

$$\alpha = (12)(36)(47)(56) ,$$
$$\beta = (13)(24)(58)(67) ,$$
$$A_0 = (14)(26)(35)(78) ,$$
$$B_1 = (16)(28)(34)(57) ,$$
$$A_1 = (18)(27)(36)(45) .$$

Since $\mathcal{Q}_0$ contains α, β, A_0, A_1 and $\mathcal{Q}_1$ contains α, β, A_1, B_1, and

$\mathcal{Q}_7$ contains α, β, A_0, B_1, the 5-set is disjoint and anti-commutative.

However, among the 8 lines not covered by the spreads of the 5-set are

the lines (46), (48), (68). Since no two disjoint spreads can cover these

three lines, the 5-set is a maximal disjoint set. In particular, then, we

have an illustration of the case $N(5) = 0$.

This proves (xv) and completes our discussion of the complete

8-point.

R. H. Bruck

From (ii), (v), (x), (xi), we have

$$p(3) \geq 30 + 960 + 630 + 420 = 2040 ,$$

with equality only if there are no further types of packing of D.

IV. <u>m = 4.</u> Here

(3.9)
$$v = 10, \ b = 45, \ k = 2, \ r = 0, \ \lambda = 1,$$
$$v' = 28, \ b' = 63, \ k' = 4, \ r' = 9, \ \lambda' = 1,$$
$$m = 4, \ n = 8 .$$

There is indeed a projective plane of order $n = 8$, (a unique one) and it has at least one complete oval. We need only refer to §2 for the case $n = 2m = 8$, $e = 2$, $s = 4$, $t = 2$. What we want to show here is that the packings studied for the previous case $m = 3$ are related in a significant way to the problem of constructing a complement for a complete 10-point.

Let D be a complete 10-point and let D' be a complement of D, so that D, D' determine the unique projective plane π of order 8, and D appears as a complete oval of π. Let L be a line of π which meets D and hence contains two distinct points of D. That is, L is a line of D. Let D_0 be the complete 8-point obtained from D by removing the common points of L and D. Then each point P of L which is not in D determines a spread of D_0, and the 7 spreads so obtained constitute a packing $\mathcal{P}$ of D_0. Conversely, by adding the line L of D to each of the spreads of the packing $\mathcal{P}$, we get a collection of spreads of D which specify the points of π which lie on L but are not in D.

The nature of the packing $\mathcal{P}$ of the complete 8-point D_0 will depend both on how D is imbedded as a complete oval in π and on how

R. H. Bruck

the line L has been chosen. Let us take, for the imbedding, the example from §3. Thus

$$D = \mathcal{C} \cup \{K\}$$

where $\mathcal{C} = \mathcal{C}(4)$ is a "conic" and K is its knot. Since t = 2, the group of all collineations of π which map D upon itself fixes K and has a (normal) subgroup of index e = 3 which is sharply triply-transitive on the 9 elements of $\mathcal{C}$.

First let the line L contain K and hence have exactly one point in common with $\mathcal{C}$. Then the subgroup, $G(\mathcal{P})$, of S_8 which maps $\mathcal{P}$ upon itself has a subgroup (of collineations of π) which is (at least) doubly transitive on D_0 and has order $8 \cdot 7 \cdot 3$. Therefore

$$|G(\mathcal{P})| = 8 \cdot 7 \cdot 3 \cdot N$$

for some positive integer N. Thus $\mathcal{P}$ must be projective. And, incidentally, N = 8. Note that L is acting as the plane at infinity for the affine 3-space determined by $\mathcal{P}$.

Next let L contain two distinct points of $\mathcal{C}$. In this case, D_0 contains K. Nevertheless, $G(\mathcal{P})$ has a subgroup (of collineations of π) which fixes K, is transitive on the remaining 7 points of D_0, and has order $7 \cdot 3$. Thus

$$|G(\mathcal{P})| = 7 \cdot 3 \cdot N$$

for some positive integer N. Hence either (a) $\mathcal{P}$ is projective and N = 64 or (b) $\mathcal{P}$ is non-projective cyclic and N = 2. By a closer examination, one may verify that the correct alternative is (b).

These remarks should suggest that there was some point in study-

R. H. Bruck

ing the spreads and packings of the complete 8-point, even though we know in advance that it has no complement.

And we can go further. No one knows whether there exists a projective plane of order 10, even if we insist on one which has a complete oval (a complete 12-point). A first step in deciding this question should involve a careful study   of the packings of the complete 10-point.

4. (k = 3.) <u>Steiner triple systems.</u> A Steiner triple system is a 1-design with k = 3. Such a design D satisfies

$$v \equiv 1 \text{ or } 3 \mod 6 .$$

If D is to have a complement, D', then 3/v and hence

(4. 1) $v = 6m + 3, \ b = (3m + 1)(2m + 1) ,$

              $k = 3, \ r = 3m + 1, \ \lambda = 1,$

and

(4. 1') $v' = m(3m - 2), \ b' = (3m + 1)(3m - 2) ,$

              $k' = m, \ r' = 3m + 1, \ \lambda' = 1 ,$

(4. 2) $n = 3m .$

where (since k' > 1) m is an integer greater than 1.

Our basic question can be phrased as follows: For what positive integers m does there exist a projective plane π of order n = 3m with a pointset D of order v = 3m + 2 imbedded in π such that D is a Steiner triple-system of order v in the sense that each line of π contains either no point of D or exactly 3 distinct points of D?

R. H. Bruck

For $m = 2$, $n = 6$, there is no plane whatsoever. Nevertheless, it might be of interest to examine the spreads and packings of the various Steiner triple systems of order 15. These systems of order 15 have been listed several times (there are 80 distinct isomorphism classes) and could certainly be examined.

For $m = 3$, the problem becomes more challenging. On the one hand, there are at least 3 isomorphic planes of order 9 (the Desarguesian plane and two planes originally constructed by Veblen and Wedderburn, belonging to the classes of Hall planes and Hughes planes). On the other hand, the number of non-isomorphic Steiner triple systems of order 21 is very large (Doyen estimates that there are more than two million). Hence it is unclear how best to attack the problem.

One answer (not relevant to $m = 3$) is to limit attention to nice classes of triple systems. For example, there are just two classes of Steiner triple systems for which the automorphism group is transitive on ordered triangles. Consider the system of points and lines of a finite-dimensional projective or affine space over $GF(q)$. For the one class, we use projective space with $q = 2$. For the other class, we use affine space with $q = 3$.

However, it seems more natural at this point to generalize the problem by not insisting on $q = 2$ or 3.

R. H. Bruck

5. (k = q or q + 1.) <u>Affine and projective spaces.</u> Let q be any prime-power, and let s,t be positive integers.

By the <u>affine design</u> AD(s,q) we mean the system consisting of the points and lines (used as varieties and blocks, respectively) of the s-dimensional affine space AG(s,q) over the finite field GF(q).

By the <u>projective design</u> PD(s,q) we mean the design of points and lines of the s-dimensional projective space, PG(s,q), over GF(q).

Both of these designs have $\lambda = 1$.

The divisibility conditions $k|v$, necessary for the existence of a complementary design, is satisfied by AD(s,q) for all s but requires s to be odd in the case of PD(s,q). Thus we take s = 2t + 1 in the following comparison:

For AD(2t + 1,q): $k = q$, $k' = (q^{2t} - 1)/(q - 1)$.

For PD(2t + 1,q): $k = q + 1$, $k' = q(q^{2t} - 1)/(q^2 - 1)$.

For either: $n = q(q + 1)(q^{2t} - 1)/(q^2 - 1)$.

Clearly if, for some positive integer t, one of AG(2t + 1,q), PD(2t + 1,q) has a complement, then there exists a projective plane of the order n given above. For t > 1 this clearly poses two distinct problems, since the parameters of PD(2t + 1,q) are not those of a complement to AD(2t + 1,q). The case of t = 1 is less clear:

(5.1)   <u>Parameters of PD(3,q)</u>:

$v = (q^2 + 1)(q + 1)$, $b = (q^2 + 1)(q^2 + q + 1)$,

$k = q + 1$, $r = q^2 + q + 1$, $\lambda = 1$.

R. H. Bruck

(5.2) Parameters of AD(3,q):

$v' = q^3$, $b' = q^2(q^2 + q + 1)$,

$k' = q$, $r' = q^2 + q + 1$, $\lambda' = 1$.

(5.3) Common order: $n = q(q + 1)$.

That is, each of AD(3,q), PD(3,q) has the parameters of a complement of the other. Nevertheless, there is no reason to assume that if one has a complement, the latter is isomorphic to the other. (I made such an assumption in my 1963 Saskatoon lectures, but now regard it as unwise.)

In the rest of the paper, we restrict attention to PD(3,q), thus leaving a great deal untouched.

6. Regular spreads of PG(3,q). Constructions of packings. In Bruck [3] I exploited a connection between regular spreads of PG(3,q) and Miquelian inversive planes (represented in terms of the points and projective sub-lines of order q in the projective line over $GF(q^2)$.) But Miquelian planes have other representations, for example in terms of the points, lines and circles of an affine plane over GF(q). I now wish to exhibit a construction in which a regular spread of PG(3,q) is related in a natural geometric way to $q^2 - q$ Miquelian inversive planes, two in the "projective line" representation and the rest (if $q > 2$) in the "affine plane" representation. The construction then can be used to obtain an alternative approach to the packings of R. H. Denniston [14]. I should add that it was Denniston's work, together with my desire to avoid his use of the Klein quadric, which led me to the present results.

I shall repeat just enough from Bruck [3] to make the present proofs understandable.

R. H. Bruck

We may imbed $\Sigma = PG(3, q)$ in $\Sigma' = PG(3, q^2)$ so that Σ consists
of subsets of the points, lines and planes of Σ' under the incidence relation of Σ. Then Σ' has precisely two types of points, those in Σ and
those not in Σ. Similarly, Σ' has precisely two types of planes. With
a line L of Σ' is associated a non-negative integer n such that L
contains exactly n points of Σ and lies in exactly n planes of Σ; and
n can be $q + 1$, 1 or 0. If a line L of Σ' contains no point of Σ then
through each of the $q^2 + 1$ points of L there passes a unique line of Σ.
The set, $\mathcal{S}(L)$, of these lines of Σ is a spread of Σ (that is, of
PD(3, q)) and is regular in the following sense: For each line M of Σ
which is not in $\mathcal{S}(L)$, the set, $\mathcal{R}$, consisting of the $q + 1$ lines of
$\mathcal{S}(L)$ which meet M, is a regulus of Σ. That is, $\mathcal{R}$ is a set of $q + 1$
disjoint lines of Σ such that every line of Σ which meets 3 disjoint
lines of $\mathcal{R}$ meets every line of $\mathcal{R}$. We shall have to repeat the proof
that, for every regular spread $\mathcal{S}$ of Σ, there are precisely two distinct
lines L of Σ' of type $n = 0$ such that $\mathcal{S} = \mathcal{S}(L)$.

Theorem 6.1. Let $\mathcal{S}$ be a regular spread of $\Sigma = PG(3, q)$. Let
A be a line of Σ. Let Σ be imbedded in $\Sigma' = PG(3, q^2)$ and let π be
one of the $q^2 - q$ planes of Σ' which contains A but is not in Σ. Then:

(i) π can be chosen in 2 distinct ways so that the q^2 distinct
lines of $\mathcal{S}$, distinct from A, meet π in q^2 distinct points of a line
L of π of type $n = 0$. The remaining point, Ω, of L is on A (but
not in Σ.) The points Ω, Ω', obtained in this way corresponding to the
two choices of π, are distinct and may be called the circular points at
infinity corresponding to $\mathcal{S}$ and A.

R. H. Bruck

(ii) If $q > 2$, π can be chosen in

$$q^2 - q - 2 = (q - 2)(q + 1)$$

ways so that the q^2 distinct lines of $\mathcal{J}$, distinct from A, meet π in the q^2 distinct points of an affine (Baer) subplane, β, of π, of order q, with A as the line at infinity of β. The $q + 1$ points of A which act as the points at infinity for β are not in Σ and are distinct from the circular points Ω, Ω'. Each regulus of $\mathcal{J}$ which contains A determines (and is determined by) q distinct points of β lying on a line of β. Each regulus of $\mathcal{J}$ which does not contain A determines (and is determined by) a circle of β; that is, a conic of π which meets A in the circular points Ω, Ω' and contains exactly $q + 1$ distinct points of β.

Proof. We represent $\Sigma = PG(3, q)$ in terms of a 4-dimensional vector space V over $GF(q)$; points, lines and planes being represented by 1-dimensional, 2-dimensional and 3-dimensional subspaces, respectively, of V over $GF(q)$. Once we have chosen a basis

$$(6.1) \qquad \{ e_1, e_2, e_1', e_2' \}$$

of V over $GF(q)$, we shall be especially interested in the line

$$(6.2) \qquad L(\infty) = \langle e_1, e_2 \rangle$$

and the lines

$$(6.3) \qquad L(X) = \langle x_{11}e_1 + x_{12}e_2 + e_1', \; x_{21}e_1 + x_{22}e_2 + e_2' \rangle \; ,$$

one for every 2×2 matrix X over $GF(q)$. The line $L(X)$ is skew to $L(\infty)$, and every line skew to $L(\infty)$ has form $L(X)$ for a unique X. Two

R. H. Bruck

lines $L(X), L(Y)$ are skew if and only if the matrix $X - Y$ is nonsingular.

We may assume without loss of generality that the basis (6.1) has been chosen so that

(6.4) $$A = L(\infty) ,$$
$$\mathcal{S} = A \cup \{L(aI + bU) | a, b \in GF(q)\} ,$$

where I denotes the 2×2 identity matrix and U a (fixed) irreducible 2×2 matrix over $GF(q)$.

Let λ be a characteristic root of U. Then λ is in $GF(q^2)$ but not in $GF(q)$, and λ^q is a second characteristic root, distinct from λ. Clearly $u_{12} \neq 0$, else U would be reducible. Hence the equations

(6.5) $$u_{12}\sigma + u_{22} = \lambda, \quad u_{12}\sigma^q + u_{22} = \lambda^q$$

determine a pair of distinct conjugate elements, σ, σ^q, of $GF(q^2)$. We note that

(6.6) $$u_{12}\sigma^2 - (u_{11} - u_{22})\sigma - u_{21} = 0 ,$$

and similarly with σ replaced by σ^q.

Next we imbed $\Sigma = PG(3, q)$ in $\Sigma' = PG(3, q^2)$ by extending V to a vector space V' over $GF(q^2)$ having the same basis (6.1) as V. In addition (to complete the imbedding) we identify a subspace of V' over $GF(q^2)$ which happens to have a basis of elements of V with the corresponding subspace of V over $GF(q)$.

For each element, ρ, of $GF(q^2)$ we define the plane $\pi(\rho)$ of Σ' by

(6.7) $$\pi(\rho) = \langle e_1, e_2, \rho e_1' + e_2' \rangle .$$

A necessary and sufficient condition that $\pi(\rho)$ should not be in Σ is that ρ be not in $GF(q)$. Henceforth we assume:

$$(6.8) \qquad \rho \in GF(q^2), \quad \rho \notin GF(q) .$$

Then

$$(6.9) \qquad A = L(\infty) \subset \pi(\rho) ;$$

and <u>the only points of Σ contained in $\pi(\rho)$ are those points of Σ contained in A.</u>

To keep things clear, let $M_2(q)$, $M_2(q^2)$ represent the set of all 2×2 matrices with elements in $GF(q)$, $GF(q^2)$ respectively.

For each X in $M_2(q)$, the line $L(X)$ of Σ meets $\pi(\rho)$ in the point

$$(6.10) \qquad P(\rho;X) = \langle v(\rho;X) + \rho e_1' + e_2' \rangle$$

where $v(\rho;X)$ is the vector in V' defined by

$$(6.11) \qquad v(\rho;X) = (\rho x_{11} + x_{21})e_1 + (\rho x_{12} + x_{22})e_2 .$$

In particular, the "point at infinity"

$$(6.12) \qquad P(\infty, \rho;X) = \langle v(\rho;X) \rangle , \quad X \neq 0 ,$$

is a point of A for each $X \in M_2(q)$, $X \neq 0$, and may or may not be in Σ. Specifically (for $X \in M_2(q), X \neq 0$) <u>$P(\infty;\rho,X)$ is in Σ precisely when X is a singular matrix.</u>

If X, Y are in $M_2(q)$, $X \neq Y$, the vector subspace of V' over $GF(q^2)$ containing the points $P(\rho;X)$, $P(\rho;Y)$ is the same as the vector subspace containing $P(\rho;X)$ and the point (at infinity), $P(\infty, \rho;Y - X)$.

R. H. Bruck

Thus $\underline{P(\rho;X), P(\rho;Y)}$ are distinct and the line

(6.13) $\qquad P(\rho;X) + P(\rho;Y) = P(\rho;X) + P(\infty, \rho;Y - X)$

has type $\underline{n = 0}$ precisely when the lines $\underline{L(X), L(Y)}$ of $\underline{\Sigma}$ are skew.

Now we are ready to study the q^2 points

(6.14) $\qquad P(\rho;aI + bU), \quad a, b \in GF(q)$.

These are q^2 distinct points of $\pi(\rho)$. They will lie on a single line of $\pi(\rho)$ precisely when the vectors

(6.15) $\qquad v(\rho;I) = \qquad \rho e_1 + \qquad e_2 ,$
$\qquad v(\rho;U) = (\rho u_{11} + u_{21})e_1 + (\rho u_{12} + u_{22})e_2 ,$

are linearly dependent over $GF(q^2)$. Since σ and σ^q satisfy (6.6), the linearly dependent case corresponds to $\rho = \sigma$ or σ^q. The relevant points at infinity corresponding to $\pi(\sigma)$, $\pi(\sigma^q)$ are

(6.16) $\qquad \Omega = P(\infty), \sigma;I) = \langle \sigma e_1 + e_2 \rangle ,$
$\qquad \Omega' = P(\infty, \sigma^q;I) = \langle \sigma^q e_1 + e_2 \rangle .$

These are points of A which are distinct and not in Σ. This proves (i).

For the proof of (ii) we strengthen (6.8) to

(6.17) $\qquad \rho \in GF(q^2), \quad \rho \notin GF(q), \quad \rho \neq \sigma, \sigma^q .$

For (6.17) to yield anything, we require $q > 2$. First we must establish the affine subplane $\beta(\rho)$ consisting of the q^2 points (6.14) and the lines joining them. It is easy to see that a line of $\pi(\rho)$ joining two distinct points (6.14) contains exactly q distinct points (6.14). Moreover, the set of all points of A lying on such lines consists of the $q + 1$ distinct points

(6.18) $P(\infty, \rho; I)$ and $P(\infty, \rho; aI + U)$, $a \in GF(q)$.

None of these is in Σ, and (especially in view of (6.23) below) each is distinct from Ω and Ω'. This is enough to show that $\beta(q)$ is an affine subplane of order q with A as its line at infinity.

Next we appeal to the known fact that, if X_1, X_2 are elements of $M_2(q)$ and $X_2 - X_1$ is non-singular, the unique regulus of Σ containing the three skew lines

$$A = L(\infty), \quad L(X_1), \quad L(X_2)$$

consists of A and the lines $L(X)$ with X of form

$$X = X_1 + a(X_2 - X_1)$$
$$= (1 - a)X_1 + aX_2, \quad a \in GF(q) .$$

Applying this fact to the lines other than A of $\mathcal{S}$, we see that the regulus of $\mathcal{S}$ which contain A correspond in a one-to-one manner to the lines of $\beta(q)$.

Finally, consider any matrix K in $M_2(q)$ such that the line $L(K)$ is not in $\mathcal{S}$. Equivalently, the matrix

$$aI + bU - K, \quad a, b \in GF(q) ,$$

is nonzero for all choices of a, b in $GF(q)$. Since $\mathcal{S}$ is a spread, $L(K)$ must meet exactly $q + 1$ distinct lines of $\mathcal{S}$. Since $\mathcal{S}$ is regular, these $q + 1$ lines must constitute a regulus, $\mathcal{R}$, of $\mathcal{S}$. Since $L(K)$ is skew to $A = L(\infty)$, $\mathcal{R}$ does not contain A. Hence $\mathcal{R}$

consists of $q + 1$ distinct lines of form $L(aI + bU)$, $a, b \in GF(q)$, such that

(6. 19) $|aI + bU - K| = 0, \quad a, b \in GF(q)$.

Conversely, to every regulus $\mathcal{R}$ of $\mathcal{S}$ which does not contain A, there corresponds at least one matrix K in $M_2(q)$ such that $L(K)$ is not in $\mathcal{S}$ and such that $\mathcal{R}$ consists of the lines $L(aI + bU)$, $a, b \in GF(q)$ subject to the (non-homogeneous) determinantal equation (6. 19). It follows also that $\mathcal{R}$ corresponds to the set of all points

(6. 20) $P(\rho; aI + bU), \quad a, b \in GF(q)$,

of $\pi(\rho)$ subject to (6. 19).

The equation (6. 19) also determines a point-set of $\pi(\rho)$ consisting of all points

(6. 21) $(x, y, z) = \langle xv(\rho; I) + yv(\rho; U) + z(\rho e_1' + e_2') \rangle$

subject to the homogeneous determinantal equation

(6. 22) $|xI + yU - zK| = 0$.

Here x, y, z are elements of $GF(q^2)$, not all zero. The points (6. 20) are the points (6. 21) with x, y in $GF(q)$ and $z = 1$. Equation (6. 22) is, of course, the equation of a conic in $\pi(\rho)$. The conic meets the line A in the "points at infinity" obtained by taking $z = 0$, $y = -1$ in (6. 22). This yields

$$|xI - U| = 0 \, ,$$

whence either

$$x = \lambda = \sigma u_{12} + u_{22}$$

or $x = \lambda^q$. Hence the "points at infinity" of the conic are

(6. 22)
$$(\lambda, -1, 0) = \langle v(\rho; \lambda I - U) \rangle \, ,$$
$$(\lambda^q, -1, 0) = \langle v(\rho; \lambda^q I - U) \rangle \, .$$

Since λ, λ^q are not in $GF(q)$, neither point is a point at infinity for β.

In (6. 22), we have, for the first time, used matrices, $\lambda I - U$ and $\lambda^q I - U$, which are in $M_2(q^2)$ but not in $M_2(q)$. We shall show in a moment that

(6. 23)
$$(\lambda, -1, 0) = \langle \sigma^q e_1 + e_2 \rangle = \Omega' \, ,$$
$$(\lambda^q, -1, 0) = \langle \sigma e_1 + e_2 \rangle = \Omega \, .$$

As soon as (6. 23) is established, the proof of (ii) and Theorem 6. 1 will be complete. Indeed, let P_1, P_2, P_3 be three distinct points of the affine subplane $\beta(\rho)$. The corresponding lines L_1, L_2, L_3 of $\mathscr{A}$ are mutually skew and hence determine a unique regulus, $\mathscr{R}$, of $\mathscr{A}$. Assume, further, that P_1, P_2, P_3 do not lie on a line of $\beta(\rho)$. Then $\mathscr{R}$ does not contain A. Hence $\mathscr{R}$ determines a unique conic of π which contains $q + 1$ distinct points of $\beta(\rho)$, including P_1, P_2, P_3, and meets A in the points (6. 22). If (6. 23) holds, the conic contains P_1, P_2, P_3, Ω, Ω' and is determined by these 5 points.

To prove (6. 23), we note that

$$v(\rho; \lambda I - U) = s e_1 + t e_2$$

where, since $\lambda = \sigma u_{12} + u_{22}$,

R. H. Bruck

$$t = \rho(-u_{12}) + (\sigma u_{12} + u_{22}) - u_{22}$$
$$= (\sigma - \rho)u_{12} \neq 0$$

and

$$s = \rho(\sigma u_{12} + u_{22} - u_{11}) + (-u_{21})$$
$$= (\rho - \sigma)(\sigma u_{12} + u_{22} - u_{11}) + 0 \quad \text{by (6.6)}$$
$$= (\rho - \sigma)(-\sigma^q u_{12})$$
$$= t\sigma^q \, .$$

Hence

$$v(\rho; \lambda I - U) = t(\sigma^q e_1 + e_2)$$

and

$$(\lambda, -1, 0) = \langle \, v(\rho; \lambda I - U) \, \rangle$$
$$= \langle \, \sigma^q e_1 + e_2 \, \rangle = \Omega' \, .$$

Similarly for the second formula of (6.23).

This completes the proof of Theorem 6.1.

The lemma which follows is a corollary of Theorem 6.1.

Lemma 6.2. Let $\mathscr{J}$ be a regular spread of $\Sigma = PG(3, q)$, $q > 2$, and let A be a line of $\mathscr{J}$. Set

(6.24) $\qquad\qquad n = q^2 + q \, ,$

and let

$$\mathscr{R}_i \, , \quad 1 \leq i \leq n \, ,$$

be the n distinct reguli of $\mathscr{J}$ which contain A. Then there exists a

R. H. Bruck

set of n regular spreads

$$\mathcal{S}_i, \; 1 \leqq i \leqq n,$$

of Σ with the following properties:

(a) For each i, the only lines common to $\mathcal{S}$ and $\mathcal{S}_i$ are the lines of $\mathcal{R}_i$.

(b) Each line of Σ which is skew to A but not in $\mathcal{S}$ is contained in exactly one of the $\mathcal{S}_i$.

Proof. We assume the situation of Theorem 6.1 and take π to be a plane of Σ' through A which is met (see (iii)) by the lines of $\mathcal{S}$ distinct from A in the points of an affine subplane, β, of π with A as its line at infinity. The corresponding projective subplane β^*, as a subplane of order q in the projective plane π of order q^2, is a Baer subplane. That is, every point of π which is not in β^* lies on a unique line of β^*.

Each of the $n = q^2 + q$ lines of β, considered in terms of the points of β alone, determines a unique regulus of $\mathcal{S}$ which contains A and, considered as a line of π which contains no point of Σ, imbeds the regulus in a regular spread of Σ. Thus we get the n reguli $\mathcal{R}_i$, each imbedded in a unique regular spread $\mathcal{S}_i$.

Now consider a line L of Σ which is skew to A. If L is common to $\mathcal{S}$ and $\mathcal{S}_i$ then L meets π in a point of β lying on the line of β which determines $\mathcal{S}_i$ and hence $\mathcal{R}_i$. Therefore L is in $\mathcal{R}_i$. This proves (a). If L is not in $\mathcal{S}$ then L meets π in a point P which is not in β^*. Hence P lies on a unique line of β^*, and this cannot be A. Therefore L lies in $\mathcal{S}_i$ for a unique choice of i. This proves Lemma 6.2.

R. H. Bruck

As Denniston [14] remarks, the 240 packings of PG(3,2) (q = 2)
are well-known. They will not be considered here. The theorem which
follows, although it asserts more than Denniston [14], was stated in
present form by Denniston at the 1972 conference in Bressanone.

Theorem 6.3. (Denniston) If $q > 2$, $\Sigma = PG(3,q)$ has at least two
inequivalent types of packing formed from one regular spread and $q^2 + q$
spreads which are subregular of index 1.

Proof. Consider the regular spreads $\mathscr{S}$ and $\mathscr{S}_i$ of Lemma 6.2.
For each $i = 1, 2, \ldots, n$, let $\mathscr{S}'_i$ be obtained from $\mathscr{S}_i$ by replacing
$\mathcal{R}_i$ by the opposite regulus, $\mathcal{R}'_i$. Thus $\mathscr{S}'_i$ is a subregular spread
of index 1 (for terminology, see Bruck[3].) We now demonstrate that the
collection

$$\mathcal{P}: \; \mathscr{S}, \; \mathscr{S}'_1, \; \mathscr{S}'_2, \ldots, \; \mathscr{S}'_t$$

of $1 + t = q^2 + q + 1$ spreads is a packing. First, the line A is in $\mathscr{S}$
but meets every line of every $\mathcal{R}'_i$ and hence is in no $\mathscr{S}'_i$. Next consider
a line L of Σ distinct from A. If L meets A then L meets the lines
of a unique regulus $\mathcal{R}_i$ of $\mathscr{S}$ containing A, and hence is in $\mathcal{R}'_i$
and $\mathscr{S}'_i$ for a unique i. If L is skew to A then there are two cases.
If L is in $\mathscr{S}$, suppose that L is also in $\mathscr{S}'_i$ for some i. Since
$\mathscr{S}$, $\mathscr{S}_i$ have only the lines of $\mathcal{R}_i$ in common, and since no line of
$\mathcal{R}_i$ is in $\mathscr{S}'_i$, the line L must be in $\mathcal{R}'_i$. But then L meets A, a
contradiction. Thus (for L skew to A) if L is in $\mathscr{S}$ then L is in no
$\mathscr{S}'_i$. Finally, let L be not in $\mathscr{S}$. Then L is skew to A (whence L
meets no $\mathcal{R}_i$ and thus is in no $\mathcal{R}'_i$) and L is in $\mathscr{S}_i$ for a unique
i. Therefore L is in a unique $\mathscr{S}'_i$.

R. H. Bruck

Now we have proved that $\mathcal{P}$ is a packing of Σ. We note (the case i = 1) that $\mathcal{R}_1$ is in $\mathcal{S}$ (and $\mathcal{S}_1$) and $\mathcal{R}_1'$ is in $\mathcal{S}_1'$. Let $\mathcal{S}'$ be the spread obtained from $\mathcal{S}$ by reversing $\mathcal{R}_1$. Thus $\mathcal{R}_1$ is in $\mathcal{S}_1$ and $\mathcal{R}_1'$ is in $\mathcal{S}'$. Hence, if $\mathcal{P}^*$ is the result of replacing $\mathcal{S}$, $\mathcal{S}_1'$ in $\mathcal{P}$ by $\mathcal{S}'$, $\mathcal{S}_1$ respectively, $\mathcal{P}^*$ is also a packing. Moreover, $\mathcal{P}$ and $\mathcal{P}^*$ each comprise 1 regular spread and $t = q^2 + q$ subregular spreads of index 1. However, the regular spreads, $\mathcal{S}$, $\mathcal{S}_1$ are differently related to the

$$t - 1 = (q - 2)(q + 1) + 1 > q + 1$$

subregular spreads

$$\mathcal{S}_2', \dots, \mathcal{S}_t',$$

since the reguli $\mathcal{R}_2, \dots, \mathcal{R}_t$ of $\mathcal{S}$ are opposite to the reguli

$$\mathcal{R}_i' \text{ of } \mathcal{S}_i' \quad (1 \leq i \leq t)$$

whereas, for each such i, no regulus of $\mathcal{S}_1$ can be opposite to a regulus of $\mathcal{S}_i'$. Therefore $\mathcal{P}$, $\mathcal{P}^*$ are inequivalent.

Next I wish to treat briefly a different type of collection of spreads. Recall that if PD(3, q) has a complement (in the sense of §1) then, in particular, there exists at least one collection, $\mathcal{C}_s$, of spreads of Σ = PG(3, q) such that every pair of skew lines of Σ lies in exactly one member of $\mathcal{C}_s$. It occurred to me to wonder whether such a $\mathcal{C}_s$ could be the orbit of a spread $\mathcal{S}$ under a suitable group, $\mathcal{H}$, of collineations of Σ.

This brought me, in time, to the concept of an <u>admissible pair</u>

R. H. Bruck

$(\mathscr{S}, \mathscr{G})$, subject to the following conditions:

(i) $\mathscr{S}$ is a spread and $\mathscr{G}$ is a group of collineations of $\Sigma = PG(3, q)$.

(ii) $\mathscr{G}$ is transitive on the set of all ordered pairs of skew lines of Σ.

(iii) If θ is in $\mathscr{G}$ and if the spreads $\mathscr{S}$, $\mathscr{S}\theta$ have more than one common line, then $\mathscr{S} = \mathscr{S}\theta$.

Given such a pair $(\mathscr{S}, \mathscr{G})$, or given a pair subject to (i), (iii) and to (ii) weakened by dropping the word "ordered", then the orbit of $\mathscr{S}$ under $\mathscr{G}$ has the above-mentioned property that every pair of skew lines of Σ lies in exactly one member of the orbit.

Hearing of the concept of an admissible pair $(\mathscr{S}, \mathscr{G})$ by word of mouth, various geometers expressed deep skepticism as to existence but no one offered a proof of non-existence (at least, so far as I know). Finally, in 1970, I turned the problem over to my Ph. D. student Sung-chi Lin, who will receive his Ph. D. degree on this topic in the spring of 1973. Some of his results may be summarized as follows:

An admissible pair $(\mathscr{S}, \mathscr{G})$ does not exist in any of the following cases:

(a) $\mathscr{S}$ is regular.

(b) $\mathscr{S}$ is subregular of index less than $(q - 1)/2$.

(c) q is an odd power of 2 and the subgroup of $\mathscr{G}$ mapping $\mathscr{S}$ upon $\mathscr{S}$ is the corresponding Suzuki group.

There are other results but a complete proof of non-existence has not been obtained. On the other hand, I know of no example of an admissible pair.

R. H. Bruck

Another topic which seems to be relevant here concerns a method of constructing an affine plane α of order q^2 in terms of certain spreads of $\Sigma = PG(3,q)$.

We choose a line A of Σ and, for the point-set, $\mathcal{P}$, of α, we take the set of all lines of Σ which are skew to A. The line-set of α will be partitioned into two disjoint sets, $\mathcal{L}$ and $\mathcal{C}$, where $\mathcal{L}$ is the set of all planes of Σ which do not contain A and where $\mathcal{C}$ is a collection of spreads of Σ containing A, with properties to be discussed. The containing relation will serve to decide whether a line L in $\mathcal{P}$ (as a point of the proposed affine plane α) is incident with a plane in $\mathcal{L}$ or spread in $\mathcal{C}$ (as a line of α).

A plane π in $\mathcal{L}$ meets A in a unique point Q and hence contains precisely q^2 distinct lines in $\mathcal{L}$, every two of which intersect. A spread $\mathcal{S}$ in $\mathcal{C}$ contains A together with precisely q^2 distinct lines in $\mathcal{L}$, every two of which are skew. Hence α, if it exists, must have order q^2.

Let L, M be distinct line of Σ, both skew to A. If L, M intersect, they lie in no spread of Σ and they lie in a unique plane, LM, of Σ. Moreover, LM does not contain A. Thus LM must be the unique line of α containing L and M. If L, M are skew then they lie in no plane of Σ. Hence, if α is to be a plane, $\mathcal{C}$ must contain a unique spread containing L and M. Thus, at the very least, $\mathcal{C}$ must satisfy:

(I) $\mathcal{C}$ is a collection of spreads of $\Sigma = PG(3,q)$ each containing a fixed line A of Σ.

(II) To every pair of lines L, M of Σ such that A, L, M are mutually skew, there corresponds one and only one spread in $\mathcal{C}$ which

R. H. Bruck

<u>contains L, M.</u>

Now we wish to show that (if $\mathcal{C}$ satisfies (I), (II)) then α, with points, lines and incidence as defined above, is indeed an affine plane (of order q^2). We know that every two distinct points of α lie on a unique line of α. Next we wish to discuss the parallel classes of lines of α. Let π be in $\mathcal{L}$, so that π is a plane of Σ meeting A in a unique point, say Q, of Σ. If π' is a plane of Σ not containing Q (and hence not containing A) then the line of intersection of π, π' in Σ is skew to A; hence π, π' are intersecting lines of α. If $\mathcal{S}$ is a spread in $\mathcal{C}$ then (by a general property of spreads of finite 3-space) $\mathcal{S}$ contains a unique line L of π. Since π does not contain A, then L is distinct from A. Since $\mathcal{S}$ is a spread containing A, L, with $A \neq L$, then L is skew to A. Hence π, $\mathcal{S}$ are intersecting lines of α. We now see that $\mathcal{L}$ (the set of all planes of Σ not containing A) is partitioned into $q + 1$ parallel classes of lines of α. Indeed, for each of the $q + 1$ distinct points Q of A, there is a parallel class consisting of the q^2 planes of Σ which contain Q but not A. We also see what we hope to prove about $\mathcal{C}$, namely:

(III) $\underline{\mathcal{C}}$ <u>is partitioned into</u> $\underline{q^2 - q}$ <u>disjoint classes of</u> $\underline{q^2}$ <u>spreads</u> <u>each. Two distinct spreads in</u> $\mathcal{C}$ <u>belong to the same class if and only if</u> <u>they have only the line A in common.</u>

To show that (I), (II) imply (III) we need some counting arguments. Let Q be a point of A. If L is a line of Σ skew to A there is a unique plane π which contains L and Q and hence does not contain A. If π is a plane containing Q but not A, then π contains exactly q^2 lines skew to A; and there are q^2 such planes π. Hence the above-mentioned

R. H. Bruck

parallel classes of lines in $\mathcal{L}$ behave as they should. In addition, the number of distinct lines of Σ skew to A is

$$|\mathcal{P}| = q^4 .$$

Next, if L is a line of Σ skew to A, we need the number of lines M of Σ skew to both of A, L. Given such an M, let π be the plane of Σ containing Q and M. Then π contains neither A nor L. Hence π meets A in Q and L in a point R, and therefore contains precisely

$$(q^2 + q + 1) - 1 - 2q = q^2 - q$$

lines skew to A and L. Letting π range over the $q^2 - 1$ planes of Σ which contain Q but neither A nor L we see that the number of lines of Σ skew to A and L is

$$(q^2 - 1)(q^2 - q) .$$

Now, by using these results, we see that the total number of spreads in $\mathcal{C}$ is

$$|\mathcal{C}| = q^4 \cdot (q^2 - 1)(q^2 - q)/q^2(q^2 - 1)$$
$$= q^2(q^2 - q)$$

and that, for every line L skew to A, the number of members of $\mathcal{C}$ containing L is

$$(q^2 - 1)(q^2 - q)/(q^2 - 1) = q^2 - q .$$

Finally, let $\mathcal{S}$ be a spread in $\mathcal{C}$ and let L be a line of Σ which is skew to A and not in $\mathcal{S}$. We prove (III) by showing that there is one and only one spread $\mathcal{S}'$ in $\mathcal{C}$ which contains L and has only A in

R. H. Bruck

common with $\mathcal{S}$. First we observe that L meets $\mathcal{S}$ in q + 1 distinct lines of Σ, necessarily distinct from A. None of these can occur with L in a spread. There remain precisely

$$(q^2 + 1) - 1 - (q + 1) = q^2 - q - 1$$

lines of $\mathcal{S}$ which are skew to L and A. Each of these is contained with L in a spread in $\mathcal{C}$, and no two in the same spread. This accounts for

$$q^2 - q - 1$$

distinct spreads in $\mathcal{C}$ which contain L. There remains in $\mathcal{C}$ exactly one additional spread $\mathcal{S}$ ' which contains L, and $\mathcal{S}$ ' has only A in common with $\mathcal{S}$. To sum up: (I), (II) imply (III).

In the presence of (I), (II), (III), it should be obvious that α is an affine plane of order q^2.

In my 1963 Saskatoon lectures (Bruck [2]) I touched on the above (potential) construction of an affine plane α of order q^2 and gave one example, in which α was Desarguesian. Later, in the fall of 1963, T. G. Ostrom sent me (in a private letter) ample evidence that "every known plane of order q^2" had such a representation. So far as I know, Ostrom's results have never reached print in this form. I cannot presume to publish them here but what I will do is connect the topic with a realm of ideas very close to Ostrom's present work.

As in the situation of Theorem 6.1, we imbed $\Sigma = PG(3, q)$ in $\Sigma' = PG(3, q^2)$ and choose a plane π of Σ' which contains the given line A but is not in Σ. Let π_0 be the Desarguesian affine plane of order q^2 obtained from π by deleting the line A and its points. For every pair

P, Q of distinct points of π_0, let PQ be the line of π_0 through P and Q, considered as a set of q^2 distinct points of π_0. By the point at infinity on PQ we shall mean the point of intersection of the corresponding line of π with A.

There is a natural correspondence between the set of points of π_0 and the set $\mathcal{P}$ of lines Σ skew to A. Indeed, each point of π_0 lies on a unique member of $\mathcal{P}$ and each member of $\mathcal{P}$ meets π in a unique point of π_0. We shall use this correspondence.

As Ostrom has frequently remarked, any affine plane of order q^2 can be regarded as having the same point-set as the Desarguesian plane π_0. Consider any affine plane π_0' on the same point-set as π_0 and, for every two distinct points P, Q of π_0, let (PQ)' be the line of π_0' through P, Q, considered as a set of q^2 distinct points of π_0. The following condition turns out to be necessary and sufficient that the above natural correspondence be an isomorphism of π_0' upon one of the planes α defined in terms of a collection $\mathcal{C}$, subject to (I), (II):

(#) If P, Q are distinct points of π_0 such that the point at infinity on PQ is in Σ, then (PQ)' = PQ.

This condition is perhaps deceptively simple, in that we are assuming that π_0' is a plane of order q^2. Thus, if R, S are two distinct points of π_0 and if (RS)' $\neq$ RS then, for every two distinct points P, Q of (RS)', the point of infinity on PQ is not in Σ. For, if, for some such P, Q, the point at infinity of PQ were in Σ, we would have

$$R, S \in (RS)' = (PQ)' = PQ$$

and hence

R. H. Bruck

$$RS = PQ \ ,$$

in contradiction to the necessary assumption that the point of infinity on
RS is not in Σ.

Now let us examine (#). First let P, Q be two distinct points of
π_0 such that the point of infinity on PQ, say the point R of A, is in Σ.
Then the line PR of π contains exactly one point, namely R, of Σ and
hence lies in exactly one plane of Σ. The latter is in $\mathscr{L}$ (see our description of α). Conversely, each member of $\mathscr{L}$ meets π in a line
containing exactly one point of Σ, namely a point of A. Thus the common
line $(PQ)' = PQ$ of π_0 and π_0' is represented by a unique member of
$\mathscr{L}$. Next assume that the point, R, of infinity on PQ is not in R.
Consider the line $(PQ)'$ of π_0', which may or may not be equal to PQ.
For every two distinct points S, T of $(PQ)'$, the point of infinity on ST
is not in Σ. This means that the line of π (not just π_0) containing
S, T has no points of Σ. In particular, the lines of Σ through S, T are
skew to each other and to A. Hence the q^2 lines of Σ corresponding to
the points of $(PQ)'$, together with A, form a spread $\mathscr{A}(P, Q)$.

It should now be clear that, given π_0' subject to (#), we can construct $\mathscr{C}$ subject to (I), (II) so that α is isomorphic to π_0'. Conversely, given α and hence given $\mathscr{C}$ subject to (I), (II), we can construct π_0' isomorphic to α.

For the Desarguesian case, we simply take $(PQ)' = PQ$ for every
pair of distinct points P, Q of π_0'. In general the point-sets $(PQ)'$ may
be more complicated; some, for example, may be affine subplanes of π_0
of order q (with the line A as line at infinity and with no points at
infinity in Σ).

R. H. Bruck

One last remark: Instead of attempting to construct π_0' or all the spreads in $\mathcal{C}$, we might look for new ways of constructing a single spread of Σ in terms of the circle of ideas surrounding condition (#).

7. <u>Translation planes of order q^2</u>. In this section I wish to indicate (without proof) some improvements in that theory of classification of translation planes which was developed in Bruck [3].

To begin with, we need to know that there exists a construction process which assigns to each spread $\mathcal{S}$ of PG(3,q) a translation plane $\pi(\mathcal{S})$ of order q^2. Next we should know that, when q is a prime, every translation plane of order q^2 has such a representation $\pi(\mathcal{S})$. On the other hand, when q is not a prime, a translation plane π has a representation $\pi(\mathcal{S})$ precisely when π is, in a precise sense, d-dimensional over GF(q) where d = 1 or 2. (d = 1 only for the Desarguesian plane.)

I have in mind the construction process developed by Bose. (See Bruck and Bose [7].) However, Segre [12] developed an equivalent process at about the same time and - to our intense surprise - we found that we had been anticipated by André [1], in slightly different language, but in a paper which must have been read by all of us.

Whatever the construction process, one would like to be assured that one can characterize the translation planes $\pi(\mathcal{S})$ in terms of the spreads $\mathcal{S}$. Thus the following theorem is crucial:

<u>Theorem 7. 1.</u> (Lüneburg.) Let $\mathcal{S}$, $\mathcal{S}'$ be spreads of PG(3,q). A necessary and sufficient condition that the translation planes $\pi(\mathcal{S})$, $\pi(\mathcal{S}')$ be isomorphic is that the spreads $\mathcal{S}$, $\mathcal{S}'$ be equivalent; that is,

R. H. Bruck

that there exist a collineation of PG(3,q) which maps $\mathscr{S}$ upon $\mathscr{S}'$.

Theorem 7.1 calls for a classification of spreads of PG(3,q) into equivalence classes. Such a classification has been completed only for q = 2,3. For q = 2 there is one equivalence class (all spreads are regular). For q = 3 there are two classes; one consists of the regular spreads and the other of the subregular spreads of index 1. For q > 3 many examples are known but the only extensive classification is that of the subregular spreads, which I shall now discuss.

Recall that a regulus, $\mathscr{R}$, of PG(3,q), is a set of q + 1 (distinct and mutually) skew lines of PG(3,q) with the property that every line of PG(3,q) which meets 3 distinct lines of $\mathscr{R}$ meets all of them. The set, $\mathscr{R}'$, consisting of the q + 1 distinct transversals to a regulus $\mathscr{R}$, is also a regulus, and $(\mathscr{R}')' = \mathscr{R}$. Moreover, $\mathscr{R}$ and $\mathscr{R}'$ cover the same points. Hence, if $\mathscr{S}$ is a spread containing a regulus, $\mathscr{R}$, and if $\mathscr{S}'$ is the line-set obtained from $\mathscr{S}$ by reversing $\mathscr{R}$ (that is, by replacing $\mathscr{R}$ by $\mathscr{R}'$) then $\mathscr{S}'$ is also a spread. We say that $\mathscr{S}'$ has been obtained from $\mathscr{S}$ by a simple reversal. Conversely, $\mathscr{S}$ can be obtained from $\mathscr{S}'$ by a simple reversal.

Let us say that two spreads $\mathscr{S}$, $\mathscr{S}'$ are reversal-equivalent provided that $\mathscr{S}'$ can be obtained from $\mathscr{S}$ by a finite (possibly empty) sequence of simple reversals. Then reversal-equivalence is an equivalence relation. True, it is only of interest in connection with spreads which contain at least one regulus and (answering a conjecture in Bruck and Bose [7] in the negative) there are now many examples of spreads which contain no reguli. Equally true, other ways of generating spreads, apart from using simple reversals, are known. Nevertheless, reversal-equivalence is very useful and provides many interesting unsolved

R. H. Bruck

problems.

Recall, next, that if A, B, C are three (distinct and mutually) skew lines of PG(3,q), there exists one and only one regulus $\mathcal{R}(A, B, C)$ of PG(3,q) which contains A, B, C. A spread, $\mathcal{S}$, of PG(3,q) is called regular provided that, for every three distinct (and hence skew) lines A, B, C of $\mathcal{S}$, every line of the regulus $\mathcal{R}(A, B, C)$ is in $\mathcal{S}$. As is well-known, the spread $\mathcal{S}$ is regular if and only if the translation plane $\pi(\mathcal{S})$ is Desarguesian.

A spread, $\mathcal{S}$, of PG(3,q) is said to be subregular if it is reversal-equivalent to at least one regular spread, and to have index k (where k is a non-negative integer) if it can be obtained from some regular spread by a sequence of k simple reversals but cannot be obtained from any regular spread by a shorter sequence of simple reversals. As a consequence, if, for some integer $k \geq 0$, there exists a subregular spread $\mathcal{S}$ of index k, then every spread $\mathcal{S}'$ equivalent to $\mathcal{S}$ (under the collineation group of PG(3,q)) also is subregular of index k. The subregular spreads of index 0 are the regular spreads and form a single equivalence class. There are no other spreads for q = 2. For q > 2, subregular spreads of index 1 exist and form a single equivalence class, corresponding to the Hall planes. For q = 3 there are no other spreads, and for q = 4 there are no other subregular spreads. However, for $q \geq 5$, subregular spreads of index 2 exist but do not form a single equivalence class. (In Bruck [3], tables are given which show the facts for $q \leq 11$ and $k \leq 5$, along with additional information.)

A serious effort at classifying subregular spreads rest upon the fact that a regular spread $\mathcal{S}$ of PG(3,q) may be regarded as a Miquelian inversive plane M(q). It is easy to show the following about

R. H. Bruck

a regular spread $\mathcal{S}$:

(i) $\mathcal{S}$ contains exactly $q^2 + 1$ distinct lines and $q(q^2 + 1)$ distinct reguli. Each regulus consists of $q + 1$ distinct lines of $\mathcal{S}$.

(ii)   Three distinct lines A, B, C of $\mathcal{S}$ lie in one and only one regulus of $\mathcal{S}$.

(iii)   Let $\mathcal{R}$ be a regulus of $\mathcal{S}$, and A, D be lines of $\mathcal{S}$ such that A is in $\mathcal{R}$ but D is not in $\mathcal{R}$. Then there exists one and only one regulus $\mathcal{R}_1$ of $\mathcal{S}$ which contains D and has only A in common with $\mathcal{R}$.

(iv)   If $\mathcal{R}$ is a regulus of $\mathcal{S}$, there exists one and only one permutation $\theta = \theta(\mathcal{R})$ of the lines of $\mathcal{S}$ which maps reguli of $\mathcal{S}$ upon reguli of $\mathcal{S}$, fixes every line of $\mathcal{R}$, but is not the identity permutation. Moreover, θ has order 2 and permutes the $q^2 - q$ lines of $\mathcal{S} - \mathcal{R}$ in cycles of length 2.

We call $\theta = \theta(\mathcal{R})$ inversion with respect to the regulus $\mathcal{R}$. And we say that two distinct lines A, B of $\mathcal{S}$ are __conjugate__ with respect to $\mathcal{R}$ provided that $\theta = \theta(\mathcal{R})$ interchanges A, B. Two further properties are of frequent use:

(v)   If $\mathcal{R}_1$, $\mathcal{R}_2$ are disjoint reguli of $\mathcal{S}$, then $\mathcal{R}_1$, $\mathcal{R}_2$ have one and only one common pair A, B of conjugate lines. If $\mathcal{R}_1$, $\mathcal{R}_2$ are distinct but not disjoint reguli of $\mathcal{S}$, then $\mathcal{R}_1$, $\mathcal{R}_2$ have no common pair of conjugate lines.

(vi)   If A, B are distinct lines of $\mathcal{S}$, there exist exactly $q - 1$ distinct reguli of $\mathcal{S}$ having A, B as a conjugate pair. These $q - 1$ reguli are disjoint, and each of the $q^2 - 1$ lines of $\mathcal{S}$ distinct from A, B lies in exactly one of these reguli.

One fact seems worth recording in connection with (iv). Let $\mathcal{R}$

R. H. Bruck

be a regulus of the regular spread $\mathcal{S}$, and let $\mathcal{Q}$ be the unique quadric of PG(3,q) with $\mathcal{R}$, $\mathcal{R}'$ as its two sets of rulings. Finally, let ϕ be the unique polarity of PG(3,q) which maps each point of $\mathcal{Q}$ upon its tangent plane. Then ϕ induces $\theta = \theta(\mathcal{R})$ upon the lines of $\mathcal{S}$.

Now we need the concepts of linear and non-linear sets (and of complete sets) of disjoint reguli of a regular spread. Let

(7. 1) $$\mathcal{R}_1, \mathcal{R}_2, \ldots, \mathcal{R}_k$$

be a set of $k \geq 1$ disjoint reguli of a regular spread $\mathcal{S}$. We say that (7. 1) is <u>linear</u> if and only if there exists a pair of distinct lines A, B of $\mathcal{S}$ which are conjugate with respect to each of the reguli $\mathcal{R}_i$. By (iv), (v), (7. 1) is certainly linear if $k = 1$ or 2. However (for $q \geq 5$) there are non-linear sets of three or more disjoint reguli. We say that (7. 1) is <u>complete</u> if $k = q - 1$; in this case, there exist exactly two distinct lines A, B which are not in any of the $\mathcal{R}_i$. We may call A, B the <u>carriers</u> of the complete set of disjoint circles. By (vi), any two distinct lines of $\mathcal{S}$ can serve as the carriers of a (unique) complete linear set of distinct reguli of $\mathcal{S}$. (We leave aside, for the moment, the question as to existence of complete non-linear sets of disjoint reguli.)

The following theorem is proved in Bruck [3]:

Theorem 7. 2. Let $\mathcal{S}$ be a regular spread of PG(3,q), let $\mathcal{C}$ be a complete linear set of disjoint reguli of $\mathcal{S}$, and let $\mathcal{S}'$ be a spread obtained from $\mathcal{S}$ by reversing some t of the reguli in $\mathcal{C}$, where

(7. 2) $$0 \leq t \leq q - 1.$$

Define

R. H. Bruck

(7.3) $k = \text{Min}(t, q - 1 - t)$.

Then $\mathcal{S}'$ is subregular of index k.

Note that, in particular, if $\mathcal{S}^*$ is obtained from $\mathcal{S}$ by reversing all of the reguli in $\mathcal{C}$, then $\mathcal{S}^*$ is regular. It is also true that the set, $\mathcal{C}^*$, of reversed reguli, is a complete linear set of disjoint circles of $\mathcal{S}^*$. And $\mathcal{S}'$ is obtained both from $\mathcal{S}$ by reversing t reguli of $\mathcal{C}$ and from $\mathcal{S}^*$ by reversing q - 1 - t reguli of $\mathcal{C}^*$. When q is odd and

$$k = t = q - 1 - t = (q - 1)/2 ,$$

this dual origin of $\mathcal{S}'$ causes some trouble (which can, however be overcome).

The spreads, $\mathcal{S}'$, which arise as in Theorem 7.2 correspond to the so-called André planes (including the Hall planes for k = 1). The André planes are completely classified in Bruck [3]. The classification of these planes (and the more general "subregular" planes) rests heavily on the following:

Theorem 7.3. Let $\mathcal{S}'$ be a spread of PG(3,q). Let k be an integer satisfying

(7.4) $0 \leqq k < (q - 1)/2$.

Then:

(i)   If $\mathcal{S}'$ is subregular of index k, there exists a unique regular spread $\mathcal{S}$ of PG(3,q) and a unique set (7.1) of k disjoint reguli of PG(3,q) such that $\mathcal{S}'$ is obtained from $\mathcal{S}$ by reversing all

R. H. Bruck

of the reguli in (7. 1). Conversely,

(ii)   If $\mathcal{S}$ ' is obtained from a regular spread $\mathcal{S}$ by reversing
a set (7. 1) of k disjoint reguli of $\mathcal{S}$, then $\mathcal{S}$ ' is subregular of
index k.

Clearly Theorem 7. 3, in so far as it applies, reduces the study of
subregular spreads of index k to a study of a fixed regular spread
and the determination of the equivalence classes of sets of k disjoint
reguli of $\mathcal{S}$ under the group of all collineations of PG(3, q) which map
$\mathcal{S}$ upon $\mathcal{S}$. But, for index k not less than (q - 1)/2, there are
other problems. These have been resolved by my student William Orr.

In his Ph. D. thesis (Madison, Wisconsin, May 1973) Orr proves
the following:

Theorem 7. 4.   (Orr.)   Let $\mathcal{S}$ be a regular spread of PG(3, q), and
let (7. 1) be a non-linear set of k disjoint reguli of $\mathcal{S}$. (Hence
$3 \leq k \leq q - 1$.) Let $\mathcal{S}$ ' be the spread obtained from $\mathcal{S}$ by reversing
each of the reguli in (7. 1). Then the only reguli in $\mathcal{S}$ ' are (a) the k
reversed reguli $\mathcal{R}_i'$ and (b) the reguli of $\mathcal{S}$ disjoint from the reguli
$\mathcal{R}_i$ of (7. 1).

I had proved (Bruck [3]) an earlier version of Theorem 7. 4 in which
the hypothesis of non-linearity was dropped but k was required to satisfy
the inequality

$$2k \leq q - 4 .$$

This result was needed for Theorem 7. 3. Using Theorems 7. 2, 7. 3, 7. 4,
we can now state:

R. H. Bruck

Theorem 7.5. Assume $q > 3$. Let $\mathcal{S}'$ be a subregular spread of $PG(3,q)$ of positive index. Then one of the following holds:

(I)   There exists a unique regular spread $\mathcal{S}$, and a unique set (7.1) of $k > 0$ disjoint reguli of $\mathcal{S}$ such that either (a) (7.1) is non-linear or (b) (7.1) is linear and $2k < q - 1$, with the property that $\mathcal{S}'$ is obtained from $\mathcal{S}$ by reversing the reguli in (7.1). In this case, $\mathcal{S}'$ has index k.

(II) q is odd and there exists a unique pair $\mathcal{S}, \mathcal{S}^*$ of distinct regular spreads of $PG(3,q)$ such that $\mathcal{S}^*$ is obtained from $\mathcal{S}$ by reversing a complete linear set $\mathcal{C}$ of disjoint reguli of $\mathcal{S}$ (resulting in a complete linear set $\mathcal{C}^*$ of disjoint reguli of $\mathcal{S}^*$) and $\mathcal{S}'$ can be obtained either from $\mathcal{S}$ by reversing a unique set of $(q-1)/2$ disjoint reguli of $\mathcal{C}$ or from $\mathcal{S}^*$ by reversing a unique set of $(q-1)/2$ disjoint reguli of $\mathcal{C}^*$. In this case, $\mathcal{S}'$ has index $(q-1)/2$.

As previously remarked, for $q = 2$ every spread is regular. For $q = 3$, the subregular spread of index $(q-1)/2 = 1$ is truly exceptional, since it can be obtained from 10 distinct regular spreads by reversing a regulus. These remarks explain why we require $q > 3$ in Theorem 7.5.

If we ignore the case (II) of Theorem 7.5, which has been studied completely in Bruck [3], Theorem 7.5 tells us that the problem of classifying subregular spreads (and the corresponding translations planes) reduces to the classification of sets of disjoint circles of a finite Miquelian inversive plane $M(q)$.

But this fact does not end the problem; indeed, there is much to be done. Let us give a crude summary of some of the work of Bruck [3] in the case that

R. H. Bruck

$$q = p^e$$

is a large prime-power, where p is a prime and $e \geq 1$ a positive integer. Consider only the subregular spreads of index 3, corresponding to a triple of disjoint reguli. The number of equivalence classes corresponding to a linear triple is asymptotic to

$$q^3/6e \; ,$$

and the number of equivalence classes corresponding to a non-linear triple is asymptotic to

$$q^3/48e \; .$$

Conclusion: For q large, there are enormously many inequivalent sub-regular spreads of $PG(3, q)$ and non-isomorphic translation planes of order q^2.

As a first unsolved problem, there is the classification of non-linear sets of 4 disjoint reguli. Another natural topic is the classification of maximal non-linear sets of disjoint reguli. The best result on the latter problem is brand new:

Theorem 7.6. (Thas, Orr.) Every complete set of reguli of a regular spread of $PG(3, q)$ is linear.

At the time of my Bressanone lectures, Thas announced that he had proved Theorem 7.6 for q even. More generally, he had proved the case q even of the following theorem:

Theorem 7.6*. (Thas, Orr.) Every flock of an ovoidal inversive plane over a finite field $GF(q)$ is linear.

R. H. Bruck

For q even, Theorem 7.6[*] is more general than Theorem 7.6, since Miquelian inversive planes are ovoidal but not every ovoidal inversive plane is Miquelian. On the other hand, for q odd, Theorem 7.6[*] is equivalent to Theorem 7.6, since the ovoidal inversive planes coincide with the Miquelian inversive planes. Thas's proof for q even does not work for q odd, and Orr's proof for q odd (which will appear in his thesis) does not work for q even.

I hope that Thas's paper will appear in this volume just after the present paper.

I will close with a result of Orr for the case $q = 9$ which appears quite exceptional. A complete (hence linear) set of disjoint circles of $M(q) = M(9)$ would consist of 8 circles. Orr discovered a non-linear set of 7 disjoint circles, one of which is orthogonal to the other 6. The collineation group mapping the 7 circles upon themselves fixes the orthogonal circle and is doubly transitive on the other 6 circles.

8. The higher dimensional cases. For t a positive integer, let $\mathcal{S}$ be a spread (this time, of $(t-1)$-dimensional projective subspaces) of $PG(2t - 1, q)$. Just as in section 7 (which corresponds to $t = 2$) to each such $\mathcal{S}$ there corresponds a translation plane $\pi(\mathcal{S})$ of order q^t. However, for $t > 2$, the theory, to my mind, is much more rudimentary.

For $t = 2$, the reguli of a doubly ruled quadric in $PG(3, q)$ played an important role. I looked for a suitable analog of this quadric in $PG(2t - 1, q)$, and found it for the case that t is an odd prime. The corresponding surfaces are algebraic surfaces of degree t in $PG(2t - 1, q)$ and are ruled by t distinct classes of $(t-1)$-dimensional projective subspaces. These surfaces (or hypersurfaces, as one auditor insisted) seem

R. H. Bruck

to be quite new. See Bruck [4].

A regulus of $PG(2t - 1, q)$ is a set $\mathcal{R}$ of $q + 1$ disjoint (t-1)-dimensional projective subspaces having the property that every line of $PG(2t - 1, q)$ which meets 3 distinct spaces of $\mathcal{R}$ meets all of them. Any set of three disjoint (t-1)-spaces lies in one and only one regulus. A spread $\mathcal{S}$ (of (t-1)-spaces) is called regular provided that, for every 3 distinct spaces of $\mathcal{S}$, $\mathcal{S}$ contains the corresponding regulus. ($\mathcal{S}$ is regular precisely when $\pi(\mathcal{S})$ is Desarguesian.) If $\mathcal{S}$ is regular, the system consisting of the spaces of $\mathcal{S}$, considered as points, and the reguli of $\mathcal{S}$, considered as circles, is clearly some kind of higher-dimensional "circle geometry". I have studied this circle-geometry in some detail, again for the case that t is an odd prime. (See Bruck [5, 6].)

It should be remarked that the spreads of the present section are quite different from the spreads of lines of $PG(2t - 1, q)$, t > 2, which were discussed briefly in section 5. The first seven sections of the paper form a unified whole, but the present section is related only to section 7.

R. H. Bruck

Bibliography

1. J. André, Uber nicht-Desarguesche Ebenen mit transitiver Translations gruppe. Math. Zeitschr. <u>60</u>, 156-186 (1954).

2. R. H. Bruck, Existence problems for classes of finite projective planes. Unpublished lecture notes of the 1963 Summer Conference of the Canadian Mathematical Congress in Saskatoon.

3. R. H. Bruck, Construction problems of finite projective planes. Combinatorial Mathematics and its Applications (Proceedings of the 1967 Chapel Hill Conference.) Chapter 27, 426-514. University of North Carolina Press, 1969.

4. R. H. Bruck, Some relatively unknown ruled surfaces in projective space. Archives, Nouvelle Serie. Section des Sciences, Institut Grand-Ducal de Luxembourg, <u>34</u>, 361-376 (1970).

5. R. H. Bruck, Circle geometry in higher dimensions. (To appear in a birthday volume for R. C. Bose.)

6. R. H. Bruck, Circle geometry in higher dimensions, II. Geometrial Dedicata (to appear).

7. R. H. Bruck and R. C. Bose, The construction of translation planes from projective spaces. Journal of Algebra, 1, 85-102 (1964).

8. R. H. Bruck and R. C. Bose, Linear representations of projective planes in projective spaces. Journal of Algebra, <u>4</u>, 117-172 (1966).

9. P. Dembowski, Finite Geometries. Ergebnisse der Mathematik und ihrer Grenzgebiete, <u>44</u>. Springer-Verlag, New York, 1968.

10. H. Lüneburg, Die Suzukigruppen und ihre Geometrien. Lecture Notes in Mathematics. Springer-Verlag, Berlin, Heidelberg, New York, 1965.

R. H. Bruck

13. Esther Seiden, On a geometrical method of construction of partially balanced designs with two associate classes, Annals of Mathematical Statistics $\underline{32}$, 1177–1180 (1961).

14. R. H. F. Denniston, Some packings of projective space, Lincei Rendiconti (to appear).

J. A. Thas

Flocks of Finite Egglike Inversive Planes

J. A. Thas

1. INTRODUCTION. An ovoid 0 of the threedimensional projective space PG(3,q), $q = p^h$ and $q > 2$, is a set of $q^2 + 1$ points no three of which are collinear.

If 0 is an ovoid of PG(3,q), then an incidence structure $I(0) = (0, B, \in)$ is defined as follows:

(i)   Points are the elements of 0;

(ii)   Blocks are called circles and are the sets $P \cap 0$, where P is a plane of PG(3,q) with $|0 \cap P| > 1$.

It is straightforward to prove that I(0) is an inversive plane of order q.

We call a finite inversive plane egglike if it is isomorphic to an I(0) for some ovoid 0. We remark that every finite inversive plane of even order is egglike [3].

A flock of a finite inversive plane I is a set α of mutually disjoint circles of I such that, with the exception of precisely two points x and y, every point of I is on a (necessarily unique) circle of α. The points x,y are called the carriers of the flock.

Consider an egglike inversive plane I(0) (0 is an ovoid of PG(3,q)) and let L be a line of PG(3,q) which has no point in common with 0. Then the circles $P \cap 0$, where P is a plane containing L with $|P \cap 0| > 1$, form a flock of I(0). It was conjectured that every flock of I(0) could be obtained in that way ([1],[2]). In this paper we prove that the conjecture is true in the case that q is even.

J. A. Thas

2. LEMMA. If L is a set of $q + 1$ points in $PG(3,q)$, which has a non-empty intersection with every plane, then L is a line of $PG(3,q)$.

Proof. Let $L = \{x_1, x_2, \ldots, x_{q+1}\}$ and suppose that L is not a line. Then the line $x_1 x_2 = L'$ contains a point y, with $y \notin L$. Let L" be a line such that $y \in L"$ and $L" \cap L = \phi$. Since each plane through L" contains at least one point of L and since L" $\cap$ L = ϕ and $|L| = q + 1$, we conclude that every plane through L" contains exactly one point of L. As $|L \cap$ plane $L'L"| \geq 2$, we obtain a contradiction. So we conclude that L is a line.

3. THEOREM. If $\{C_1, C_2, \ldots, C_{q-1}\}$, $q = 2^h$ and $h > 1$, is a flock of the finite egglike inversive plane $I(0)$, then the planes of the $q - 1$ circles C_i all pass through the same line L. This line L is the intersection of the tangent planes of 0 at the carriers x, y of the flock.

Proof. First of all we remark that $0 = C_1 \cup C_2 \cup \ldots \cup C_{q-1} \cup \{x\} \cup \{y\}$. The nucleus of the circle C_i is denoted by n_i. Now we shall prove that $L = \{x, y, n_1, \ldots, n_{q-1}\}$ is a line. For that purpose we show that every plane of $PG(3,q)$ has at least one point in common with L.

a)   Every plane through x or y has a point in common with L.

b)   Let P be the tangent plane of 0 at $p \in 0$ ($p \neq x$, $p \neq y$). Through p there passes a circle of the flock, say C_i. The tangent line of C_i at p is contained in P, and so $n_i \in P$.

c)   Let P be a plane for which $P \cap 0 \in \{C_1, C_2, \ldots, C_{q-1}\}$. If $C_i = P \cap 0$, then $n_i \in P$.

d)   Finally let P be a plane for which $|P \cap 0| > 1$ and $x \notin P$, $y \notin P$, $P \cap 0 \notin \{C_1, \ldots, C_{q-1}\}$. If $P \cap 0 = C$, then $|C_i \cap C| \in \{0, 1, 2\}$. As $q + 1$ is odd there exists a C_j such that $|C_j \cap C| = 1$. There follows that C_j and C have a common tangent line T at their common point. And so $n_j \in T \subset P$, from which $n_j \in P$.

We conclude that every plane of $PG(3,q)$ has at least one point in common with the set L of order $q + 1$. From the lemma there follows that L is a line of $PG(3,q)$.

Next we remark that the polar planes of $x, y, n_1, \ldots, n_{q-1}$, with respect to the symplectic polarity π defined by 0, are the tangent plane of 0 at x, the tangent plane of 0 at y, and the planes of the circles C_i. As L is a line, these $q + 1$ planes all pass through the polar line of L with respect to π. So we conclude that the planes of the $q - 1$ circles C_i all pass through the intersection of the tangent planes of 0 at the carriers x, y of the flock.

BIBILIOGRAPHY

[1]   R. H. BRUCK, Finite geometric structures and their applications: Construction of finite planes, 2nd C. I. M. E. Session 1972 (Bressanone).

[2]   P. DEMBOWSKI, "Finite geometries", Springer-Verlag, 1968, 275 pp.

[3]   P. DEMBOWSKI and D. R. HUGHES, On finite inversive planes, J. London Math. Soc., 40 (1965), 171-182.

Prof. Dr. J. A. Thas
Seminar of Higher Geometry
University of Ghent
J. Plateaustraat 22
9000 GENT
BELGIUM

CENTRO INTERNAZIONALE MATEMATICO ESTIVO

(C. I. M. E.)

R. H. F. DENNISTON

PACKINGS OF PG(3, q)

Corso tenuto a Bressanone dal 18 al 27 Giugno 1972

PACKINGS OF PG(3,q)

by

R. H. F. Denniston

(University of Leicester)

One of the conjectures discussed in Bruck's lectures is the possibility of constructing a projective plane of order $q(q + 1)$, where q
would be some power of a prime. The "points" would be all the points,
together with some spreads of lines, in PG(3, q): the "lines" would be
all the lines of the space, together with some suitably chosen packings.

(See the lecture notes for definitions). This construction would be
impossible if there were no packings ——— but in fact many different
packings do exist.

In PG(3, 2) , packings fall into two transitivity classes under the
collineation group (Cole, Bull. Amer. Math. Soc., 28 (1922) 435-7).

These two classes are, however, interchanged by any duality (the
concept "packing" being regarded as self-dual); and so we may say that
all packings of PG(3,2) are alike. The words "alike" and "different",
in what follows, are to be understood in this strong sense.

I have to tell you that, in every PG(3, q) with q greater than
2, packings exist and are not all alike. A construction, effective for
all such values of q, is given in a note (now in the press for Rend.
Accad. Lincei) : I give a short description of it, and show one way of
adapting it to furnish a second packing.

By Klein's method, the points of PG(5, q) are used to represent the linear complexes of PG(3, q). A complicated construction in
the higher space furnishes a regular spread $\mathcal{M}$ of PG(3, q), and
$q^2 + q$ other regular spreads $\mathcal{S}_1, \mathcal{S}_2, \ldots$; and $\mathcal{M} \cap \mathcal{S}_i$, for
each i, is a regulus $\mathcal{R}_i$. We switch $\mathcal{R}_i$ out of $\mathcal{S}_i$ (see Bruck's
lectures) to give a subregular spread $\mathcal{S}_i^*$ of index 1 ——— and al-

R. H. F. Denniston

so switch $\mathcal{R}_1$ out of $\mathcal{M}$ to give a spread $\mathcal{M}^*$. My note proves that

$$\left\{ \mathcal{M}, \; \mathcal{S}_1^*, \; \mathcal{S}_2^*, \; \mathcal{S}_3^*, \ldots \right\}$$

is a packing: and a different packing is

$$\left\{ \mathcal{M}^*, \; \mathcal{S}_1^*, \; \mathcal{S}_2^*, \; \mathcal{S}_3^*, \ldots \right\} .$$

I have done some work on PG(3, 3), and found a packing that consists (like those just mentioned) of one regular spread and the rest subregular, but is more symmetrical than they are. Then there were three different packings consisting entirely of subregular spreads; and three more, in each of which all but two of the spreads were regular. So PG(3, 3), where the spreads are of only two different types, has packings of at least nine; I image that the number of types is considerably more than 9 when q = 3, and that it increases rapidly with q .

Let us define a "cyclic packing" as a set of $q^2 + q + 1$ disjoint spreads, permuted cyclically by a collineation of period $q^2 + q + 1$.

The packing of PG(3, 2) is in fact cyclic; and the concept is discussed generally by C. R. Rao (Proceedings of the 1967 Chapel Hill Conference).

There is no cyclic packing of PG(3, 4) because any collineation of period 21 has some line-orbits of length less than 21; and I should expect the same difficulty to arise whenever $q^2 + q + 1$ is not a prime. I have also found, by searching, that there is no cyclic packing of PG(3, 3). Curiously enough, there is a BIB design, with a cyclic packing, which has the same parameters as the design of points and lines in PG(3, 3) (Moore, Amer. J. Math. 18 (1896) 264-303).

At the end of this talk, I give a short arithmetical specification

R. H. F. Denniston

of six different cyclic packings of PG(3, 8). One of these consists entirely of regular spreads; and so does the packing of PG(3, 2), every spread in that space being regular. But, in PG(3, 3), I have carried out a complete search by sorting punched cards, and found that no packing with regular spreads exists. So we may take some interest in two unsolved problems: "Which three-dimensional spaces have cyclic packings?", and "Which spaces have packings that consist entirely of regular spreads?".

We could get another unsolved problem by making a definition: A "complete partial packing of deficiency d" is a set of $q^2 + q + 1 - d$ disjoint spreads in PG(3, q), such that no other spread can be found that is disjoint from all of these. Mesner (Can. J. Math. 19 (1967) 273-280) has made an analogous definition of "complete partial spread", and has set up, in connection with his definition, a substantial lower bound for the (positive) deficiency. I have found, in PG(3, 3) , a complete partial packing of deficiency 2 : on the other hand, a deficiency of 1 is easily seen to be impossible. Does the minimum deficiency increase as a function of q ?

But the important unsolved problem is: "Can Bruck's hypothetical construction ever be carried out?" About that I have nothing much to report, except for the variety of packings shown to exist, which does seem encouraging. In any set of packings that satisfied all the conditions for the construction, only a very few (if any) could be of the type given by the process that I can carry out in a general PG(3, q).

The next step, after finding a packing , P, should be to find what we might call "a transversal" od P —— a spread in which no two lines belong to same spread of P. Suppose, in fact, that Bruck's construction has been carried out, and that P is one of the packings which are used as "lines" of the plane. Then any two lines, if they are skew and belong to different spreads of P, must belong to some transversal of P. For each of the packings I have found in PG(3, 3), I

R. H. F. Denniston

have easily verified that there are not enough transversals to satisfy the condition just mentioned. In PG(3, 4), the packing given by my first construction has many transversals, but there are other conditions they do not satisfy.

In PG(3, 8), on the other hand, I have so far not managed to find any transversal to any of the packings discovered.

This must be regarded as a failure to think of a suitable method, since some transversals presumably exist; in fact, 72 seems the least unlikely order for a projective plane constructed by Bruck's method, and accordingly I hope that somebody will be able to continue my work.

Since it will be some time before the results in PG(3, 8) are published, I give here enough arithmetic to enable the packings to be reconstructed by anyone who is interested.

Let $GF(2^3)$ be constructed by adjoining to $GF(2)$ an element i such that $i^3 = i^2 + 1$. Let (x, y, z) be non-homogeneous coordinates over this field, and let g_0 be the line $y = 0$, $z = 1$. Let collineations, with the respective periods 73, 7, 9, be specified by

$$T : (x, y, z) \longrightarrow (i^3 x + i^6 y + iz, \ i^5 x + i^3 y + i^3 z, \ i^3 x + iy + i^5 z),$$

$$U : (x, y, z) \longrightarrow (ix, iy, iz),$$

$$V : (x, y, z) \longrightarrow (i^4 x^2 + i^6 y^2, \ x^2 + y^2, \ z^2).$$

Let numbers $t_0, \ldots, t_6$ be given by any row of the following table (a complete table would have twenty rows, but the packings so specified would not all be different) :

R. H. F. Denniston

	t_0	t_1	t_2	t_3	t_4	t_5	t_6
(a)	30	66	17	15	20	38	50
(b)	43	66	17	15	20	38	50
(c)	30	15	38	17	50	20	66
(d)	30	54	9	41	60	34	40
(e)	4	13	7	47	25	20	39
(f)	44	11	60	50	32	41	62

Let $\mathcal{S}$ be a set of 65 lines, two of which are the z-axis and the line at infinity on z = 0, while the others are

$$V^v \; T^{t_u} \; U^u g_0 \quad (u = 0, \ldots, 6; \; v = 0, \ldots, 8).$$

Then it will be found that $\mathcal{S}$ is a spread, and moreover that $\left\{ \mathcal{S} \, , \; T\mathcal{S} \, , \; T^2\mathcal{S} \, , \ldots, \; T^{72}\mathcal{S} \right\}$ is a packing. Row (a) of the table gives a regular spread $\mathcal{S}$, and (b) a subregular spread of index 1 . Rows (c) and (d) give two different spreads which are not subregular, but each of which contains seven disjoint reguli. Rows (e) and (f) give two different spreads, neither containing any regulus.

CENTRO INTERNAZIONALE MATEMATICO ESTIVO
(C. I. M. E.)

J. DOYEN

RECENT RESULTS ON STEINER TRIPLE SYSTEMS

Corso tenuto a Bressanone dal 18 al 27 Giugno 1972

RECENT RESULTS ON STEINER TRIPLE SYSTEMS

Jean DOYEN

Department of Mathematics

University of Brussels

1050 Brussels, Belgium

1. Introduction. A __Steiner triple system__ (briefly STS)
of __order__ v is a finite non-empty set S of v elements (called
__points__), together with a collection of subsets of S (called
__lines__) such that every line has exactly 3 points and every pair
of points is contained in exactly one line ; a Steiner triple
system of order v is sometimes denoted simply by $S(v)$. For
example, any d-dimensional finite prjective space over $GF(2)$
is an $S(2^{d+1}- 1)$ and any d-dimensional finite affine space over
$GF(3)$ is an $S(3^d)$.

An $S(v)$ is nothing else than a balanced incomplete block
design with parameters v, $b = v(v - 1)/6$, $r = (v - 1)/2$, $k = 3$,
$\lambda = 1$. As b and r must be integers, it follows that $v \equiv 1$ or 3
(mod 6). Kirkman [13] proved in 1847 that this necessary condition of existence is also sufficient : in other words, there
is an $S(v)$ for every $v \equiv 1$ or 3 (mod 6).

2. Constructions. Roughly speaking, the methods of cons-
tion of STS are of two types : the __direct constructions__, in
which an STS is constructed directly from an algebraic structure, and the __recursive constructions__, in which an STS is
obtained from a collection of "smaller" STS. We shall give an
example of each type:

(A) Let G be a finite multiplicative abelian group of odd
order 2t + 1. Take as points the 6t + 3 elements of the set

J. Doyen

$S = G \times \{0,1,2\}$ and as lines the following subsets of S :

(i) $\{(x,0),(x,1),(x,2)\}$ for every $x \in G$

(ii) $\{(x,0),(y,0),(z,1)\}$, $\{(x,1),(y,1),(z,2)\}$,

$\{(x,2),(y,2),(z,0)\}$ for every x, y, $z \in G$ such that

$x \neq y$ and $xy = z^2$.

This direct construction, which is essentially due to Bose[1], yields a Steiner triple system of order $6t + 3$ on the set S. We shall denote it by S_G. It can be shown[6] that two triple systems S_{G_1}, S_{G_2} are isomorphic if and only if the groups G_1, G_2 are isomorphic.

(B) Let $S_1 = \{x_1,\dots,x_m\}$ be a triple system of order m and $S_2 = \{y_1,\dots,y_n\}$ a triple system of order n. Take as points the mn elements of the product set $S = S_1 \times S_2$ and as lines the following subsets of S :

(i) $\{(x_i,y_r),(x_i,y_s),(x_i,y_t)\}$ for every point x_i of S_1 and every line $\{y_r,y_s,y_t\}$ of S_2

(ii) $\{(x_i,y_r),(x_j,y_r),(x_k,y_r)\}$ for every line $\{x_i,x_j,x_k\}$ of S_1 and every point y_r of S_2

(iii) $\{(x_i,y_r),(x_j,y_s),(x_k,y_t)\}$ for every line $\{x_i,x_j,x_k\}$ of S_1 and every line $\{y_r,y_s,y_t\}$ of S_2.

This recursive construction yields a Steiner triple system of order mn on the set S. We shall call this system the <u>direct</u> <u>product</u> of S_1,S_2 and denote it simply by $S_1 \times S_2$.

H. Werner [25] proved that Steiner triple systems have a unique factorization property with respect to this product : the decomposition of a given STS into irreducible factors is unique up to isomorphism of the factors and up to their order.

3. <u>Isomorphisms</u>. It is still an open problem to determine the number $N(v)$ of pairwise non isomorphic STS of a given order v. It is easy to check that $N(1) = N(3) = N(7) = N(9) = 1$. De Pasquale [4] proved in 1899 that $N(13) = 2$. One of the S(13)

.J. Doyen

can be constructed by taking as points the elements of the
additive group Z_{13} of integers modulo 13 and as lines all
subsets of Z_{13} of the form $\{x, x + 1, x + 4\}$, $\{x, x + 2, x + 7\}$
where $x \in Z_{13}$; the other $S(13)$ is obtained by removing the
lines $\{0,1,4\}$, $\{0,2,7\}$, $\{2,4,9\}$, $\{1,7,9\}$ from the above
system and replacing them by $\{0,1,7\}$, $\{0,2,4\}$, $\{1,4,9\}$,
$\{2,7,9\}$. The computation of $N(15)$ was first attacked (by hand)
in 1917 by Cole, Cummings and White [3]: they found $N(15) = 80$.
A computer search carried out in 1955 by Hall and Swift [12] led
to the same conclusion. The values of $N(v)$ for $v \geqslant 19$ are
still unknown.

The best estimate of this function is due to R. M. Wilson
(unpublished): he proved that

$$e^{\frac{v^2}{12}(\log v - 5)} \leqslant N(v) \leqslant e^{\frac{v^2}{6}\log v}$$

for every $v \equiv 1$ or $3 \pmod 6$. Moreover, if van der Waerden's
conjecture on the permanent of doubly stochastic matrices is
true (see for instance [20] for further details), then the 12 in
the above lower bound can be replaced by a 6.

Wilson's lower bound is not very good for small values
of v. However, applying his basic idea to the STS of order 19
and 21, one gets $N(19) \geqslant 8894$ and $N(21) \geqslant 2.10^6$.

4. Automorphisms. Let us define an incomplete Steiner
triple system as a finite non-empty set S of v points, together
with a collection of subsets of S, called lines, such that
every line has exactly 3 points and every pair of points is
contained in at most one line. C. Treash [24] has proved that
every incomplete STS can be (finitely) completed into an STS.
It is not very difficult to show [5] that given a finite abstract
group G, there exists an incomplete STS whose automorphism group
is isomorphic to G. However, the following problem is still open:

J. Doyen

Given a finite abstract group G, does there exist an STS whose automorphism group is isomorphic to G ?

Another problem has gained interest in the past few years:

Given a permutation α on a set of cardinality $v \equiv 1$ or 3 (mod 6), does there exist an STS of order v admitting α as an automorphism ? We shall denote such a system by $S_\alpha(v)$.

If α has a single cycle of length v, R. Peltesohn [17] proved in 1939 that there is an $S_\alpha(v)$ for every $v \equiv 1$ or 3 (mod 6), except for $v = 9$.

If α has just one fixed point and a cycle of length $v - 1$ on the remaining points, A. Rosa (unpublished) proved that there is an $S_\alpha(v)$ if and only if $v \equiv 3$ or 9 (mod 24).

If α is an involution with only one fixed point, a necessary and sufficient condition for the existence of an $S_\alpha(v)$ is $v \equiv 1$, 3, 9 or 19 (mod 24) (Rosa [19], Doyen[8], Teirlinck[23]).

If α is an involution with exactly 3 (necessarily colli-near) fixed points, the existence of an $S_\alpha(v)$ has been conjec-tured for every $v \equiv 1$ or 3 (mod 6), $v \neq 1$. Such systems can be constructed easily for every $v \equiv 3$ (mod 6) (take one of the systems S_G described in section 2 (A) and consider the automor-phism α of S_G defined by $\alpha(x,i) = (x^{-1},i)$ for every $x \in G$ and $i = 0,1,2$), but the problem is still unsolved for $v \equiv 1$ (mod 6).

5. Subsystems. (From now on, we shall say that a positive integer v is admissible if $v \equiv 1$ or 3 (mod 6)).

A subset S' of a Steiner triple system S is called a subsystem of S if every line of S joining two points of S' is entirely contained in S'. A subsystem S' is said to be of order v' if it has cardinality v'. Clearly, any intersection of sub-systems of S is again a subsystem of S, so that it makes sense to speak of the subsystem generated by any given subset of S.

J. Doyen

If an $S(v)$ contains a subsystem S' of order $v' \neq v$, it is easy to see that $v \geqslant 2v' + 1$. Conversely, Doyen and Wilson [10] have proved that given a Steiner triple system S' of order v' and an admissible integer $v \geqslant 2v' + 1$, there exists an $S(v)$ containing a subsystem isomorphic to S'.

Let S be a Steiner triple system of order $v \geqslant 7$. A subset T of S consisting of 3 non collinear points will be called a <u>triangle</u> of S. According to a classification introduced by L. Szamkołowicz [24], S is called

- (i) a <u>non degenerated plane</u> if every triangle of S generates the whole system (in other words, if S has no proper subsystem),

- (ii) a <u>degenerated plane</u> if one of the triangles of S generates the whole system and another one does not,

- (iii) a <u>space</u> if no triangle of S generates the whole system.

One can prove [6][7] that there is a non degenerated plane of order v for every admissible $v \geqslant 7$ and a degenerated plane of order v for every admissible $v \geqslant 15$. A. J. W. Hilton (unpublished) has constructed a space of order v for every admissible $v > 867$ but not much is known for smaller values of v. For example, the only admissible integers $v < 100$ for which a space of order v is known to exist are 15, 27, 31, 45, 49, 55, 63, 81, 91, 93 and 99. It is easy to see that there are no spaces of order 19, 21 and 25. The existence of a space of order 33 is still in doubt.

On the other hand, the only known examples of spaces in which all subsystems generated by triangles have the same order v' are those for which $v' = 7$ or 9. It is tempting to conjecture that there are no others. Note that if every triangle generates an $S(7)$, the system is necessarily a projective

space PG(d,2). However, if every triangle generates an S(9),
the system is not necessarily an affine space AG(d,3) :
M. Hall[11]has constructed an S(81) which is not isomorphic to
AG(4,3) but in which every triangle generates an S(9) ;
obviously, any direct product S(81) $\times$ S(3) $\times$ S(3) $\times$ ... $\times$ S(3)
gives an S(3^n) with the same property for every n $\geqslant$ 4.

    6. Disjoint STS. Two Steiner triple systems having the
same set of points are called disjoint if they have no line in
common.

    Let us denote by D(v) the maximum number of pairwise dis-
joint S(v) that can be constructed on a given set S of v points
(v $\geqslant$ 3). As S contains v(v - 1)(v - 2)/6 subsets   of cardinality 3 and as any S(v) has v(v - 1)/6 lines, we have immediately D(v) $\leqslant$ v - 2. Obviously, D(3) = 1. Cayley [2] and Kirkman
[14] proved in 1850 that D(7) = 2 and D(9) = 7. R. H. F.
Denniston (unpublished) has shown with a computer that D(13) =
11 , but the value of D(15) is not yet known. A lower bound
for D(v) can be found in [9].

    It has been conjectured that D(v) = v - 2 for every
admissible v $\geqslant$ 9. L. Teirlinck [22] proved recently that if
D(v) = v - 2, then D(3v) = 3v - 2. This, together with the
results mentioned above, implies that the function D(v) achie-
ves its maximum value for every v of the form 3^n or 13.3^n.

    Two disjoint Steiner triple systems S_1, S_2 are said to be
orthogonal if whenever two pairs of poits appear with the same
third point in lines of S_1, they appear with distinct third
points in lines of S_2.

    A pair of orthogonal STS of order v has been constructed
for infinitely many values of v $\equiv$ 1 (mod 6) by C. C. Lindner
and N. S.Mendelsohn [15] . On the other hand, the non existence

of a pair of orthogonal STS of order v is obvious for v = 3
and has been established by R.C. Mullin and B. Nemeth [46] for
v = 9. There was a conjecture that such a pair does not exist
for v ≢ 3 (mod 6), but Rosa [ ] has exhibited recently two
orthogonal STS of order 27. Not much is known for the other
values of v.

REFERENCES

1. R.C.Bose : On the construction of balanced incomplete
   block designs, Ann. Eugenics 9 (1939), 353-399.

2. A. Cayley : On the triadic arrangements of seven and
   fifteen things, London, Edinburgh and Dublin Philos. Mag.
   and J. Sci. (3) 37 (1850), 50-53.

3. F.N.Cole, L.D.Cummings and H.S.White : The complete enumeration of triad systems in 15 elements, Proc. Nat. Acad.
   Sci. U.S.A. 3 (1917), 197-199.

4. V. de Pasquale : Sui sistemi ternari di 13 elementi, Rend.
   R. Ist. Lombardo Sci. e Lett.(2) 32 (1899), 213-221.

5. J. Doyen : Constructions groupales d'espaces linéaires
   finis, Acad. Roy. Belg. Bull. Cl. Sci. 54 (1968), 144-156.

6. J. Doyen : Sur la structure de certains systèmes triples
   de Steiner, Math. Zeitschr. 111 (1969), 289-300.

7. J. Doyen : Systèmes triples de Steiner non engendrés par
   tous leurs triangles, Math. Zeitschr. 118 (1970), 197-206.

8. J. Doyen : A note on reverse Steiner triple systems,
   Discrete Math. 1 (1972), 315-319.

9. J. Doyen : Constructions of disjoint Steiner triple systems,
   Proc. Amer. Math. Soc. 32 (1972), 409-416.

10. J. Doyen and R.M. Wilson : Embeddings of Steiner triple
    systems, Discrete Math. (to appear).

11. M. Hall, Jr. : Automorphisms of Steiner triple systems,
    IBM J. Res. Develop. 4 (1960), 460-472.

J. Doyen

12. M. Hall, Jr. and J.D.Swift : Determination of Steiner triple systems of order 15, Math. Tables Aids Comput. 9 (1955), 146-156.

13. T.P.Kirkman : On a problem in combinations, Cambridge and Dublin Math. J. 2 (1847), 191-204.

14. T.P.Kirkman : Note on an unanswered prize question, Cambridge and Dublin Math. J. 5 (1850), 255-262.

15. C.C.Lindner and N.S.Mendelsohn : Construction of perpendicular Steiner quasigroups (to appear).

16. R.C.Mullin and E.Nemeth : On the nonexistence of orthogonal Steiner systems of order 9, Canad. Math. Bull. 13 (1970), 131-134.

17. R.Peltesohn : Eine Lösung der beiden Heffterschen Differenzenprobleme, Compositio Math. 6 (1939), 251-257.

18. A.Rosa : On reverse Steiner triple systems, Discrete Math. 2 (1972), 61-71.

19. A.Rosa : On the falsity of a conjecture on orthogonal Steiner triple systems (to appear).

20. H.J.Ryser : Combinatorial Mathematics (Carus Math. Monograph No. 14), Wiley, New York 1963.

21. L. Szamkołowicz : Sur une classification des triplets de Steiner, Rend. Accad. Naz. Lincei (8) 36 (1964), 125-128.

22. L. Teirlinck : On the maximum number of disjoint Steiner triple systems (to appear).

23. L. Teirlinck : The existence of reverse Steiner triple systems (to appear).

24. C.A.Treash : The completion of finite incomplete Steiner triple systems with applications to loop theory, J. Combinat. Theory Ser. A 10 (1971), 259-265.

25. H.Werner : A unique factorization theorem for Steiner triple systems (to appear).

CENTRO INTERNAZIONALE MATEMATICO ESTIVO
(C. I. M. E.)

H. LÜNEBURG

GRUPPEN UND ENDLICHE PROJEKTIVE EBENEN

Corso tenuto a Bressanone dal 18 al 27 Giugno 1972

Gruppen und endliche projektive Ebenen

von

Heinz Lüneburg

Ein wichtiger Teil der Theorie der endlichen projektiven Ebenen besteht in der Charakterisierung bekannter Ebenen durch ihre Kollineationsgruppen bzw. durch Untergruppen ihrer Kollineationsgruppen. Bei solchen Charakterisierungen muß man sich nun in der Regel der verschiedenartigsten Hilfsmittel bedienen. Dies soll hier am Beispiel der von O. Prohaska stammenden Charakterisierung der Hall-Ebenen detailliert beschrieben werden.

1. Projektive und affine Ebenen. Ist $\mathcal{R}$ eine Menge, deren Elemente wir __Punkte__, und $\mathcal{L}$ eine Menge, deren Elemente wir __Blöcke__ nennen, ist ferner $I \subseteq \mathcal{R} \times \mathcal{L}$, so nennen wir das Tripel $(\mathcal{R},\mathcal{L},I)$ eine __Inzidenzstruktur__. In gewissen Spezialfällen, so z. B. bei projektiven und affinen Ebenen, nennen wir die Blöcke auch __Geraden__ und ersetzen den Buchstaben $\mathcal{L}$ durch $\mathcal{G}$. Ist $P \in \mathcal{R}$ und $b \in \mathcal{L}$, so schreiben wir statt $(P, b) \in I$ bzw. $(P, b) \notin I$ meist $P I b$ bzw. $P \not I b$.

Ist $(\mathcal{R},\mathcal{G},I)$ eine Inzidenzstruktur, deren Blöcke wir Geraden nennen, so heißt $(\mathcal{R},\mathcal{G},I)$ eine __projektive Ebene__, falls $(\mathcal{R},\mathcal{G},I)$ die folgenden Bedingungen erfüllt:

1) Sind $P,Q \in \mathcal{R}$ und ist $P \neq Q$, so gibt es genau ein $g \in \mathcal{G}$ mit $P,Q \ I \ g$.

Heinz Lüneburg

2) Sind g,h $\in$ $\mathcal{G}$, so gibt es wenigstens ein P $\in$ $\mathcal{R}$ mit P I g,h.

3) Es gibt vier verschiedene Punkte in $\mathcal{R}$, von denen keine drei kollinear sind.

Dabei heißen die Punkte P,Q,R,... $\in$ $\mathcal{R}$ <u>kollinear</u>, falls es ein g $\in$ $\mathcal{G}$ gibt mit P,Q,R,... I g.

($\mathcal{R}$,$\mathcal{G}$,I) heißt <u>affine Ebene</u>, falls ($\mathcal{R}$,$\mathcal{G}$,I) die folgenden Bedingungen erfüllt:

1') Sind P,Q $\in$ $\mathcal{R}$ und ist P $\neq$ Q, so gibt es genau ein g $\in$ $\mathcal{G}$ mit P,Q I g.

2') Ist P $\in$ $\mathcal{R}$, g $\in$ $\mathcal{G}$ und ist P $\not{I}$ g, so gibt es genau ein h $\in$ $\mathcal{G}$ mit P I h und h $\cap$ g = $\emptyset$. (Dabei ist generell der Durchschnitt zweier Geraden g und h durch g $\cap$ h = $\{X | X \in \mathcal{R}, X I g,h\}$ definiert.)

3') Es gibt drei nicht kollineare Punkte in $\mathcal{R}$.

Zwei Geraden einer affinen Ebene heißen <u>parallel</u>, falls sie entweder gleich sind oder aber keinen Punkt gemeinsam haben. Sind g und h parallel, so schreiben wir g $\cap$ h. Die Parallelitätsrelation ist eine Äquivalenzrelation. Die Klassen dieser Relation nennen wir <u>Parallelenscharen.</u>

Ist Π = ($\mathcal{R}$,$\mathcal{G}$,I) eine projektive Ebene und ist g $\in$ $\mathcal{G}$, so sei

Heinz Lüneburg

$$\Pi^g = (\mathcal{R} \smallsetminus g, \mathcal{G} \smallsetminus \{g\}, I \cap ((\mathcal{R} \smallsetminus g) \times (\mathcal{G} \smallsetminus \{g\}))).$$

Dann ist Π^g eine affine Ebene und man erhält jede affine Ebene auf diese Weise.

Es sei $(\mathcal{R}, \mathcal{L}, I)$ eine Inzidenzstruktur. Ist $\mathcal{R}^d = \mathcal{L}$, $\mathcal{L}^d = \mathcal{R}$ und $I^d = \{(\ell, P) \mid (P, \ell) \in I\}$, so ist auch $(\mathcal{R}^d, \mathcal{L}^d, I^d)$ eine Inzidenzstruktur. Sie heißt die zu $(\mathcal{R}, \mathcal{L}, I)$ duale Inzidenzstruktur. Offenbar ist $(\mathcal{R}^{dd}, \mathcal{L}^{dd}, I^{dd}) = (\mathcal{R}, \mathcal{L}, I)$. Ist $\Pi = (\mathcal{R}, \mathcal{G}, I)$ eine projektive Ebene, so ist auch $\Pi^d = (\mathcal{R}^d, \mathcal{G}^d, I^d)$ eine projektive Ebene.

Die Inzidenzstruktur $(\mathcal{R}, \mathcal{L}, I)$ heißt endlich, falls $\mathcal{R}$ und $\mathcal{L}$ und damit auch I endlich sind. Ist $(\mathcal{R}, \mathcal{L}, I)$ endlich, so setzen wir

$$I_P = \{(P, \ell) \mid P \, I \, \ell\}, \quad I(P) = \{\ell \mid P \, I \, \ell\}, \quad I_\ell = \{(P, \ell) \mid P \, I \, \ell\},$$

$$I(\ell) = \{P \mid P \, I \, \ell\}.$$

Dann ist $|I_P| = |I(P)| = r_P$ und $|I_\ell| = |I(\ell)| = k_\ell$. Ferner setzen wir $v = |\mathcal{R}|$ und $b = |\mathcal{L}|$. Weil sowohl $\{I_P \mid P \in \mathcal{R}\}$ als auch $\{I_\ell \mid \ell \in \mathcal{L}\}$ eine Partition von I ist, gilt

<u>1.1. Satz.</u> Für jede endliche Inzidenzstruktur $(\mathcal{R}, \mathcal{L}, I)$ gilt

$$\sum_{P \in \mathcal{R}} r_P = \sum_{\ell \in \mathcal{L}} k_\ell.$$

Von besonderem Interesse sind die sog. taktischen Konfigurationen, das sind diejenigen endlichen Inzidenzstrukturen, für die $r_P = r$ und $k_\ell = k$ für alle P und alle ℓ gilt. Für taktische Konfigurationen hat man die Gleichung vr = bk, die unmittelbar aus 1.1 folgt.

Ist Π eine projektive Ebene und sind g und h Geraden von Π, so
gibt es stets einen Punkt P, der weder auf g noch auf h liegt.
(Dies beweist man mit Hilfe von 3).) Definiert man σ durch $X^{\sigma} =$
$= XP \cap h$ für alle X I g, so wird σ zu einer Bijektion der Menge
der Punkte auf g auf die Menge der Punkte von h. Daher liegen auf
g ebensoviele Punkte wie auf h. Es ist ebenfalls leicht einzusehen,
daß es zu jedem Punkt P und zu jeder Geraden g eine Bijektion der
Menge der Geraden durch P auf die Menge der Punkte auf g gibt. Hieraus folgt: Ist Π eine endliche projektive Ebene, so gibt es eine
natürliche Zahl n $\geqslant$ 2, so daß auf jeder Geraden von Π genau n + 1
Punkte liegen und durch jeden Punkt genau n + 1 Geraden gehen. Hieraus folgt wegen v(n + 1) = b(n + 1) weiter, daß v = b ist. Betrach-
tet man ferner einen Punkt P von Π und die Inzidenzstruktur aus
den von P verschiedenen Punkten von Π und den Geraden durch P,
so ist dies eine taktische Konfiguration mit den Parametern v' = v - 1,
k' = n, r' = 1, b' = n + 1. Daher ist v - 1 = v'r' = b'k' = (n + 1)n,
so daß v = b = n^2 + n + 1 ist. Es gilt also

1.2. Satz. Ist Π eine endliche projektive Ebene, so gibt es eine
natürliche Zahl n $\geqslant$ 2, so daß folgendes gilt:
1) Π besitzt n^2 + n + 1 Punkte.
2) Π besitzt n^2 + n + 1 Geraden.
3) Auf jeder Geraden liegen n + 1 Punkte.
4) Durch jeden Punkt gehen n + 1 Geraden.
Ist umgekehrt Π eine Inzidenzstruktur mit den Eigenschaften 1) und

3) und inzidieren zwei verschiedene Punkte von Π stets mit genau
einer Geraden, so ist Π im Falle $n \geqslant 2$ eine endliche projektive
Ebene.

Benutzt man, daß jede affine Ebene von der Form Π^g ist, so folgt

1.3. Satz. Ist A eine endliche affine Ebene, so gibt es eine natürliche Zahl $n \geqslant 2$, so daß folgendes gilt:

1) A besitzt n^2 Punkte.

2) A besitzt $n^2 + n$ Geraden.

3) Auf jeder Geraden von A liegen n Punkte.

4) Durch jeden Punkt gehen $n + 1$ Geraden.

5) Jede Parallelenschar enthält n Geraden.

Ist umgekehrt A eine Inzidenzstruktur, die 1) und 3) erfüllt, und
inzidieren zwei verschiedene Punkte von A stets mit genau einer Geraden, so ist A eine affine Ebene, falls nur $n \geqslant 2$ ist.

Die Zahl n aus 1.2 bzw. 1.3 heißt die Ordnung der projektiven bzw.
affinen Ebene.

Eine Inzidenzstruktur $(\mathcal{R}, \mathcal{G}, I)$ heißt ein Netz, falls sie die folgenden Bedingungen erfüllt:

1") Zwei verschiedene Punkte aus $\mathcal{R}$ inzidieren mit höchstens einer
Geraden aus $\mathcal{G}$.

2") Zu $P \in \mathcal{R}$ und $g \in \mathcal{G}$ mit $P \not{I} g$ gibt es genau ein $h \in \mathcal{G}$ mit

Heinz Lüneburg

P I h und h $\cap$ g = $\emptyset$.

3") Es gibt drei nicht kollineare Punkte, die zu je zweien eine Verbindungsgerade haben.

Jede affine Ebene ist also ein Netz und wie bei affinen Ebenen defi-
nieren wir auch auf den Geraden eines Netzes eine Parallelitätsrelation ∥ , die genau wie im Falle der affinen Ebenen eine Äquivalenzrelation ist. Mit Hilfe von 2") und 3") ergibt sich ferner, daß
jedes Netz eine taktische Konfiguration ist. Ist n die Punkteanzahl
auf einer Geraden, so ist n auch die Anzahl der Geraden in einer
Parallelenschar, wie aus 2") folgt. Daher ist $v = n^2$. Ferner ist
$m = b$, falls r die Anzahl der Geraden durch einen Punkt ist. r heißt
gelegentlich auch der _Grad_ des Netzes, während n wiederum die Ordnung
des Netzes ist. Es gilt offenbar $3 \leq r \leq n + 1$ und $r = n + 1$ genau
dann, wenn das Netz eine affine Ebene ist.

2. Translationsebenen. _Isomorphismen_ von Inzidenzstrukturen definiert
man in naheliegender Weise. Isomorphismen von projektiven bzw. affinen Ebenen sowie von Netzen auf sich selbst nennen wir _Kollineatio-_
nen. Isomorphismen von Π auf Π^d heißen _Dualitäten._

In der Theorie der projektiven Ebenen sind die _axialen_ und _zentralen_
Kollineationen von besonderem Interesse. Dabei heißt die Kollineation
σ axial mit der _Achse_ g, falls σ die Gerade g punktweise festläßt.
Zentral wird dual definiert. Es gilt der Satz, daß eine Kollineation
genau dann axial ist, wenn sie zentral ist. Man nennt axiale Kolli-

Heinz Lüneburg

neationen daher auch _Perspektivitäten_. Ist σ axial mit der Achse g

und dem Zentrum P, so heißt σ im Falle P $\not I$ g eine _Homologie_ oder auch

eine _Streckung_ und im Falle P I g eine _Elation_. Die Elationen mit der

Achse g bilden eine Untergruppe der Kollineationsgruppe (Beweis!),

die wir mit $T(g)$ bezeichnen. Die Bemerkung, daß die Identität die ein-

zige Perspektivität mit zwei verschiedenen Zentren (Achsen) ist, lie-

fert insbesondere, daß $T(g)$ auf der Menge der Punkte von Π^g scharf

transitiv operiert, falls sie transitiv operiert. Operiert $T(g)$ auf der

Menge der Punkte von Π^g transitiv, so heißt Π^g eine _Translations-_

ebene. $T(g)$ wird in diesem Falle auch _Translationsgruppe_ und die Ele-

mente von $T(g)$ auch _Translationen_ genannt.

Es sei Π^g eine Translationsebene. Ist P I g, so bezeichnen wir mit

$T(P,g)$ die Untergruppe aller Elationen mit dem Zentrum P und der Achse

g. Ist $\tau \in T(g)$, so ist $\tau^{-1}T(P,g)\tau = T(P^\tau,g^\tau) = T(P,g)$, so daß

$T(P,g)$ ein Normalteiler von $T(g)$ ist. Ferner ist $T(g) = \bigcup_{P\,I\,g} T(P,g)$

und $T(P,g) \cap T(Q,g) = \{1\}$, falls $P \neq Q$ ist. Ist $\sigma \in T(P,g)$ und

$\tau \in T(Q,g)$, so ist, da die Gruppen $T(X,g)$ ja Normalteiler von $T(g)$

sind, $\sigma^{-1}\tau^{-1}\sigma\tau \in T(P,g) \cap T(Q,g)$, so daß $\sigma\tau = \tau\sigma$ ist, falls

nur $P \neq Q$ ist. Gilt $\sigma,\tau \in T(P,g)$ und ist $Q \neq P$, so gibt es ein

$\varrho \in T(Q,g)$ mit $\varrho \neq 1$. Dann ist $\tau\varrho \notin T(P,g)$ und daher

$$(\sigma\tau)\varrho = \sigma(\tau\varrho) = (\tau\varrho)\sigma = \tau(\varrho\sigma) = \tau(\sigma\varrho) = (\tau\sigma)\varrho,$$

so daß auch in diesem Falle $\sigma\tau = \tau\sigma$ gilt. Daher haben wir

2.1. Satz. Ist Π^g eine Translationsebene, so ist $T(g)$ abelsch.

Heinz Lüneburg

Setze

$$K(g) = \{\eta \mid \eta \in \text{End}_Z T(g), \ T(P,g)^{\eta} \subseteq T(P,g) \text{ für alle } P \text{ I } g\}.$$

$K(g)$ heißt der <u>Kern</u> von $T(g)$. Nach ANDRÉ [1] ist $K(g)$ ein Körper, so
daß $T(g)$ ein $K(g)$-Vektorraum ist. Die $T(P,g)$ sind Unterräume dieses
Vektorraumes. Ferner gilt, daß die Gruppe der Streckungen mit der
Achse g und Zentrum $P \not\!I g$ zur multiplikativen Gruppe von $K(g)$ isomorph ist. Schließlich gilt noch, daß Π genau dann desarguessch ist,
wenn $T(g)$ als $K(g)$-Vektorraum den Rang 2 hat (ANDRÉ loc. cit.). Dabei
heißt die projektive Ebene Π desarguessch, wenn es einen Vektorraum
V vom Range 3 über einem Körper K gibt, so daß Π zu der Inzidenzstruktur isomorph ist, die aus den Unterräumen vom Range 1 als Punkten und den Unterräumen vom Range 2 als Geraden mit der Inklusion
als Inzidenzrelation besteht.

Π^g sei weiterhin eine Translationsebene. Ferner sei $T = T(g)$ und
$\pi = \{T(P,g) \mid P \text{ I } g\}$. Ist dann $\pi(T) = (T, \{X\eta \mid X \in \pi, \ \eta \in T\}, \in)$,
so ist $\pi(T)$ zu Π^g isomorph, wie man mühelos verifiziert. Ein Iso-
morphismus wird durch die Abbildung $\tau \rightarrow 0^{\tau}$ ($\tau \in T$) induziert,
wenn 0 ein festgewählter Punkt von Π^g ist. Ist andrerseits T eine
Gruppe und π eine nicht triviale <u>Partition</u> von T, dh. eine Menge
von Untergruppen von T mit den Eigenschaften:

a) Es ist $T = \bigcup_{X \in \pi} X$,

b) Sind $X,Y \in \pi$ und ist $X \neq Y$, so ist $X \cap Y = \{1\}$,

c) π enthält mindestens zwei Untergruppen,

Heinz Lüneburg

hat die Partition π ferner die Eigenschaft, daß für $X, Y \in \pi$ und
$X \neq Y$ stets $T = XY$ gilt, so ist $\pi(T)$ eine Translationsebene und die
Abbildung $*$ die $g \in T$ auf die durch $x^{g^*} = xg$ $(x \in T)$ definierte Abbildung g^* abbildet, ist ein Isomorphismus von T auf die Gruppe der Trans-
lationen von $\pi(T)$. Insbesondere folgt noch, daß T abelsch ist. Eine
nicht triviale Partition mit dieser zusätzlichen Eigenschaft, heißt
<u>Kongruenzpartition</u> von T.

<u>2.2. Satz</u> (André). <u>Ist π eine Kongruenzpartition der Gruppe T, so
ist $\pi(T)$ eine Translationsebene und T^* ist die Translationsgruppe
dieser Ebene. Auf diese Weise erhält man bis auf Isomorphie alle
Translationsebenen.</u>

Sind $X, Y \in \pi$, so gibt es ein $Z \in \pi$ mit $Z \neq X, Y$, da eine Gruppe
niemals Vereinigung zweier echter Untergruppen ist. Daher ist $T = XZ = YZ$ und folglich

$$X \cong X/X \cap Z \cong XZ/Z = YZ/Z \cong Y/Y \cap Z \cong Y.$$

Folglich sind alle <u>Komponenten</u> von π isomorph. Dies gilt sogar als
$K(T)$-Vektorraumisomorphismus, wenn $K(T)$ die Menge der Endomorphismen
von T ist, die jede einzelne Komponente von π in sich abbilden.
$K(T)$ ist natürlich zum Kern von T^* isomorph. Hat T endlichen Rang
über $K(T)$, so ist $\mathrm{Rg}\, T = 2n$ und $\mathrm{Rg}\, X = n$ für alle $X \in \pi$. Ist $L(T)$
der Verband der Unterräume von T, dh. die zu T gehörige projektive
Geometrie, so ist also π eine Überdeckung von T mit paarweise windschiefen Unterräumen des Ranges n. Ist umgekehrt π eine Überdeckung
von T mit paarweise windschiefen Unterräumen des Ranges n und ist

Heinz Lüneburg

Rg T = 2n, so ist π eine Kongruenzpartition von T. Diese projektive
Betrachtungsweise der Kongruenzpartitionen ist gelegentlich von Nut-
zen.

Ist T ein Vektorraum vom Range 2 über dem Körper K und ist π die
Menge der Unterräume vom Range 1, so ist π eine Kongruenzpartition
von T. Wegen $K \subsetneq K(T)$ ist $K = K(T)$, so daß $\pi(T)$ desarguessch ist.

3. Baer-Unterebenen. Es sei $\mathcal{R}_0 \subseteq \mathcal{R}$, $\mathcal{G}_0 \subseteq \mathcal{G}$ und $I_0 = I \cap (\mathcal{R}_0 \times \mathcal{G}_0)$.
Sind $\pi = (\mathcal{R}, \mathcal{G}, I)$ und $\pi_0 = (\mathcal{R}_0, \mathcal{G}_0, I_0)$ projektive Ebenen, so heißt
π_0 eine Unterebene von π. Ist $\pi_0 \neq \pi$, so heißt π_0 echte Unter-
ebene von π.

3.1. Satz (Bruck). Ist π_0 eine echte Unterebene der endlichen projektiven Ebene π, ist m die Ordnung von π_0 und n die Ordnung von
π, so ist entweder $m^2 = n$ oder $m^2 + m \leq n$. Dabei ist $m^2 = n$ gleich-
bedeutend mit der Aussage: Jeder Punkt von π liegt auf einer Geraden von π_0.

Beweis. Es sei P ein Punkt von π, der mit keiner Geraden von π_0
inzidiert. Dann enthält jede Gerade durch P höchstens einen Punkt
von π_0. Weil andrerseits jeder Punkt von π_0 auf einer Geraden durch
P liegt, ist $m^2 + m + 1 \leq n + 1$, so daß in diesem Falle $m^2 + m \leq n$
ist.

Jeder Punkt von π liege auf einer Geraden von π_0. Es sei P ein
Punkt von π_0. Wegen $m < n$ gibt es dann eine Gerade g durch P, die

Heinz Lüneburg

mit Π_0 nur den Punkt P gemeinsam hat. Die m^2 Geraden von Π_0, die nicht durch P gehen, schneiden g in m^2 verschiedenen Punkten, die überdies alle von P verschieden sind. Weil andrerseits jeder Punkt von g, der von P verschieden ist, auf genau einer Geraden von Π_0 liegt, die dann notwendig zu jenen m^2 Geraden gehört, folgt, daß $m^2 = n$ ist, q. e. d.

Im Falle $m^2 = n$ heißt Π_0 Baer-Unterebene von Π.

3.2. Hilfssatz. Π sei eine projektive Ebene der Ordnung m^2. Ferner seien Π_0 und Π_1 Baer-Unterebenen von Π. Ist g eine Gerade von Π_0 und Π_1, sind A und B zwei verschiedene Punkte von $\Pi_i{}^g$ (i = 0,1) und gilt schließlich $\{X | X$ Punkt von Π_0, $X \mathrel{I} g\} = \{X | X$ Punkt von Π_1, $X \mathrel{I} g\}$, so ist $\Pi_0 = \Pi_1$.

Beweis. Weil $\Pi_0 \cap \Pi_1$ (die Definition dieses Durchschnitts ist die naheliegende) vier Punkte enthält, von denen keine drei kollinear sind, ist $\Pi_0 \cap \Pi_1$ eine Unterebene, die auf Grund der weiteren Voraussetzungen mindestens die Ordnung m hat. Folglich ist $\Pi_0 = \Pi_0 \cap \Pi_1 = \Pi_1$, q. e. d.

4. Desarguessche affine Ebenen von Quadratzahlordnung. Wir beginnen mit einem Hilfssatz.

4.1. Hilfssatz. Π sei eine endliche projektive Ebene. Ferner sei (P,g) ein nicht inzidentes Punkt-Geradenpaar von Π. Schließlich sei $\varrho \subseteq I(g)$ mit $|\varrho| = q + 1$ und $\sum$ sei eine zur SL(2,q) isomor-

Heinz Lüneburg

phe Kollineationsgruppe von Π . **Gilt dann:**

a) $P^\Sigma = P$,

b) $g^\Sigma = g$,

c) **Ist** $X \in g$, **so ist** $\Sigma(X,PX) = T(X,PX) \cap \Sigma$ **transitiv auf** $g \setminus \{X\}$,
so ist $\{P\} \cup Q^\Sigma \cup g$ **die Punktmenge einer Unterebene der Ordnung**
q, **falls nur** $P \not= Q \not\mid g$ **und** $PQ \cap g \in g$ **ist. Die Geraden durch P,**
die zu dieser Unterebene gehören, sind die Geraden PX mit $X \in g$.

Beweis. Wir berechnen zunächst Q^Σ . Wegen $X = PQ \cap g \in g$ ist
$\Sigma(X,PX)$ im Stabilisator Σ_Q von Q in Σ enthalten. Daher ist
$|\Sigma_Q| \geqslant q$ und folglich $|Q^\Sigma| = |\Sigma|/|\Sigma_Q| \leqslant q(q^2 - 1)q^{-1} = q^2 - 1$.
Es sei h eine Gerade durch P mit $h \cap g \in g$. Wegen $|g| = q + 1 \geqslant 3$
gibt es zwei verschiedene Punkte X,Y mit $X,Y \in g$ und $X,Y \not\mid h$. Wegen
c) gibt es zu jedem $\varrho \in \Sigma(X,PX)$ mit $h^\varrho \not= PY$ ein $\varrho' \in \Sigma(Y,YP)$ mit
$h^{\varrho\varrho'} = h$. Es sei nun R I h und $R \in Q^\Sigma$. Ferner sei $R^{\varrho\varrho'} = R^{\sigma\sigma'}$.
Hieraus folgt, daß R^ϱ , R^σ und Y kollinear sind. Da andrerseits
X, R^ϱ und R^σ ebenfalls kollinear sind, folgt $\varrho = \sigma$ und damit
$\varrho' = \sigma'$. Also ist $|\{R^{\varrho\varrho'} \mid \varrho \in \Sigma(X,PX), h^\varrho \not= PY\}| = q - 1$.
Wegen $|g| = q + 1$ und der Transitivität von Σ auf g ist also
$|Q^\Sigma| \geqslant (q + 1)(q - 1)$, so daß $|Q^\Sigma| = q^2 - 1$ ist.

Duale Schlüsse zeigen: Ist h eine Gerade mit $P \not\mid h \not= g$ und $h \cap g$
ist ein Punkt aus g , so ist $|h^\Sigma| = q^2 - 1$. Dabei spielt g in
diesem Falle die Rolle von P und $\mathcal{Y} = \{PX \mid X \in g\}$ die Rolle von
g .

Nun ist $|\{P\} \cup Q^\Sigma \cup g| = 1 + q^2 - 1 + q + 1 = q^2 + q + 1$. Ferner

Heinz Lüneburg

ist $|\{g\} \cup h^{\mathcal{I}} \cup \mathcal{y}| = q^2 + q + 1$. Wählen wir h nun so, daß auch

noch Q I h gilt, so ist

$$\Pi_0 = (\{P\} \cup Q^{\mathcal{I}} \cup \mathcal{g}, \{g\} \cup h^{\mathcal{I}} \cup \mathcal{y}, I)$$

eine taktische Konfiguration mit den Parametern $v = b = q^2 + q + 1$,

$k = r = q + 1$ und der weiteren Eigenschaft, daß zwei Punkte von Π_0

auf höchstens einer Geraden von Π_0 liegen. Hieraus folgt, daß Π_0

eine Unterebene der Ordnung q ist, q. e. d.

4.2. Hilfssatz. **Es sei** PG(3,q) **die projektive Geometrie der Dimen-**
sion 3 über GF(q). **Ist** $\mathcal{G}$ **ein Hyperboloid in** PG(3,q), **so gibt es ge-**
nau $\frac{1}{2}q^2(q - 1)^2$ **Geraden von** PG(3,q), **die** $\mathcal{G}$ **nicht treffen.**

Beweis. Die Punkteanzahl von PG(3,q) ist $q^3 + q^2 + q + 1 = (q + 1)(q^2+1)$.

Daher ist die Anzahl der Geraden von PG(3,q) gleich $(q^2 + 1)(q^2 + q + 1)$,

da durch zwei verschiedene Punkte genau eine Gerade geht. Weil $\mathcal{G}$

ein Hyperboloid ist, enthält $\mathcal{G}$ zwei Regelscharen ρ_1 und ρ_2. Daher

ist die Anzahl der Geraden, die ganz in $\mathcal{G}$ enthalten sind, gleich

$|\rho_1| + |\rho_2| = 2(q + 1)$. Ist $P \in \mathcal{G}$, so gibt es genau eine Gerade

$g_i \in \rho_i$ mit P I g_i. In der Ebene $g_1 + g_2$ (Addition im Vektorraum)

liegen weitere q - 1 Geraden durch P. Diese haben mit $\mathcal{G}$ nur P ge-

meinsam und sind alle Geraden durch P, die mit $\mathcal{G}$ nur P gemeinsam

haben. Die restlichen q^2 Geraden durch P treffen $\mathcal{G}$ gerade in den

$(q + 1)^2 - 2q - 1 = q^2$ Punkten, die nicht auf $g_1 + g_2$ liegen. Die

Anzahl der Geraden, die mit $\mathcal{G}$ nur einen Punkt gemeinsam haben, ist

also $(q + 1)^2(q - 1)$ und die Anzahl b der Sekanten errechnet sich

aus $(q + 1)^2 q^2 = vr = bk = 2b$ zu $\frac{1}{2}q^2(q + 1)^2$, da die Punkte von $\mathcal{G}$

Heinz Lüneburg

zusammen mit den Sekanten eine taktische Konfiguration bilden. Die Anzahl der Passanten ist also

$$(q^2 + 1)(q^2 + q + 1) - 2(q + 1) - (q + 1)^2(q - 1) - \frac{1}{2}q^2(q + 1)^2 =$$

$$= \frac{1}{2}q^2(q - 1)^2,$$

q. e. d.

V sei ein Vektorraum vom Range 4 über GF(q). Ferner seien π und π' Kongruenzpartitionen von V. Ist $\pi(V)$ desarguessch (in diesem Falle nennen wir π eine desarguessche Kongruenzpartition), so ist GF(q) im Kern K von $\pi(V)$ enthalten. Weil V über GF(q) den Rang 4 hat, ist daher $[K:GF(q)] = 2$ und folglich $K \cong GF(q^2)$. Somit ist V ein Vektorraum vom Range 2 über K. Ist $\pi'(V)$ ebenfalls desarguessch, so ist V auch Vektorraum vom Rang 2 über dem Kern K' von $\pi'(V)$ und K' ist gleichfalls zu $GF(q^2)$ isomorph. Somit gibt es eine bijektive semilineare Abbildung σ des K-Vektorraumes V auf den K'-Vektorraum V. Weil π gerade aus den Unterräumen des Ranges 1 des K-Vektorraumes V und π' aus den Unterräumen des Ranges 1 des K'-Vektorraumes V besteht, ist $\pi^\sigma = \pi'$. Nun ist $GF(q) \subseteq K$ und $GF(q) \subseteq K'$. Weil ein endlicher Körper höchstens einen Körper gegebener Ordnung enthält, folgt, daß σ einen Automorphismus von GF(q) induziert. Somit ist σ eine semilineare Abbildung des GF(q)-Vektorraumes V auf sich, welche π auf π' abbildet.

Setzt man $\pi = \pi'$ in den vorstehenden Betrachtungen, so erhält man, daß die Gruppe aller semilinearen Abbildungen des K-Vektorraumes V

Heinz Lüneburg

auf sich, die π invariant lassen, zu $\Gamma L(2,q^2)$ isomorph ist. Insgesamt erhalten wir, daß die Anzahl der desarguesschen Kongruenzpartitionen gleich $|\Gamma L(4,q)|/|\Gamma L(2,q^2)|$ ist. Nun ist

$$|\Gamma L(4,q)| = |\text{Aut } GF(q)| q^6(q^4 - 1)(q^3 - 1)(q^2 - 1)(q - 1)$$

und

$$|\Gamma L(2,q^2)| = |\text{Aut } GF(q^2)| q^2(q^4 - 1)(q^2 - 1).$$

Schließlich folgt aus $|\text{Aut } GF(q^2)| = 2|\text{Aut } GF(q)|$

4.3. Hilfssatz. Die Anzahl der desarguesschen Kongruenzpartitionen des Vektorraumes V vom Range 4 über GF(q) ist $\frac{1}{2}q^4(q^3 - 1)(q - 1)$.

Sind g_1, g_2, g_3 drei paarweise windschiefe Geraden in einem Vektorraum V vom Rang 4 über dem Körper K, sind ferner h_1, h_2, h_3 drei verschiedene Transversalen von g_1, g_2, g_3 und ist $g_1 \cap h_1 = P_1$, $g_1 \cap h_2 = P_2$, $g_2 \cap h_1 = P_3$, $g_2 \cap h_2 = P_4$ und $g_3 \cap h_3 = P_5$, so ist $\{P_1,\ldots,P_5\}$ ein Rahmen, dh. eine Menge von fünf Punkten von denen keine vier in einer Ebene liegen. Weil GL(V) die Rahmen transitiv untereinander permutiert, folgt, daß GL(V) auch auf den Tripeln paarweise windschiefer Geraden transitiv operiert. Hieraus folgt wiederum, daß GL(V) auf der Menge der Regelscharen transitiv ist, da eine Regelschar ja gerade aus den sämtlichen Transversalen von drei paarweise windschiefen Geraden besteht. Projektiv liest sich das analog, nämlich: PGL(V) ist auf der Menge der Regelscharen transitiv. Da zu jeder Regelschar φ, die konjugierte Regelschar $\bar{\varphi}$ gehört, die gerade aus den sämtlichen Transversalen von φ besteht,

Heinz Lüneburg

sieht man leicht, daß $PGL(V)_\varphi$,der globale Stabilisator von φ in $PGL(V)$, zu $PGL(2,K) \times PGL(2,K)$ isomorph ist. Die Anzahl der Regelscharen ist daher im Falle $K = GF(q)$ gleich

$$|PGL(4,q)|/|PGL(2,q)|^2 = q^6(q^4 - 1)(q^3 - 1)(q^2 - 1)/(q^2 - 1)^2 q^2 .$$

Also gilt

4.4. Hilfssatz. Die Anzahl der Regelscharen in $V(4,q)$ ist gleich $q^4(q^3 - 1)(q^2 + 1)$.

4.5. Hilfssatz. Die Anzahl der Regelscharen von $V(4,q)$, die in einer desarguesschen Kongruenzpartition von $V(4,q)$ enthalten ist, ist gleich $q(q^2 + 1)$ und die Anzahl der desarguesschen Kongruenzpartitionen, die eine gegebene Regelschar enthalten, ist gleich $\frac{1}{2}q(q - 1)$.

Beweis. Es sei π eine desarguessche Kongruenzpartition von $V(4,q)$. Dann ist der Stabilisator von π in $GL(4,q)$ auf π dreifach transitiv. Hieraus folgt, daß drei verschiedene Komponenten von π stets in genau einer oder stets in keiner Regelschar enthalten sind, deren sämtliche Geraden zu π gehören. Nun operiert $SL(2,q)$ aber auf $\pi(V)$ in der Weise, wie in den Voraussetzungen von 4.1 beschrieben. Die affinen Punkte der durch $SL(2,q)$ bestimmten Baerunterebenen sind dann gerade Unterräume vom Rang 2 von $V = V(4,q)$, die darüberhinaus Transversalen von $q + 1$ der Komponenten von π sind. Folglich enthält π Regelscharen, so daß drei verschiedene Komponenten von π stets in genau einer Regelschar liegen, die ihrerseits ganz in π enthalten ist. Die 3-Teilmengen von π und die in π enthaltenen

Heinz Lüneburg

Regelscharen bilden also eine taktische Konfiguration mit $v = \binom{q^2+1}{3}$, $k = \binom{q+1}{3}$, b und $r = 1$. Wegen $vr = bk$ ist daher $b = q(q^2 + 1)$.

Wir betrachten nun die Inzidenzstruktur $(\mathcal{R}, \mathcal{C}, \subseteq)$ aus der Menge $\mathcal{R}$ der Regelscharen und der Menge $\mathcal{C}$ der desarguesschen Kongruenzen. Dann ist $(\mathcal{R}, \mathcal{C}, \subseteq)$ eine taktische Konfiguration mit den Parametern $v = q^4(q^3 - 1)(q^2 + 1)$, $b = \frac{1}{2}q^4(q^3 - 1)(q - 1)$, $k = q(q^2 + 1)$ und r. Wegen $vr = bk$ ist daher $r = \frac{1}{2}q(q - 1)$, q. e. d.

Es sei π eine desarguessche Kongruenzpartition von $V(4,q)$. Dann läßt sich π auffassen als projektive Gerade über $GF(q^2)$. Die Gruppe $PGL(2,q^2)$ operiert auf π scharf dreifach transitiv. Bildet man das Tensorprodukt $V(2,q) \otimes_{GF(q)} GF(q^2)$, so erhält man eine Einbettung der projektiven Gerade über $GF(q)$ in die projektive Gerade über $GF(q^2)$, so daß $PGL(2,q)$ auf der eingebetteten Geraden $V_0(2,q)$ in der richtigen Weise operiert. Nun enthält $PGL(2,q)$ eine zyklische Gruppe U der Ordnung $q + 1$ und diese liegt wiederum in einer zyklischen Gruppe der Ordnung $q^2 - 1$ von $PGL(2,q^2)$. Da die letztere zwei Fixpunkte hat, hat auch U zwei Fixpunkte X und Y, die jedoch beide nicht in $V_0(2,q)$ enthalten sind. Da $PGL(2,q^2)$ scharf dreifach transitiv operiert und die Transvektionen nur einen Fixpunkt haben, folgt, daß $PGL(2,q)_A$, falls A ein Punkt von $V_0(2,q)$ ist, auf den Punkten von π, die nicht in $V_0(2,q)$ liegen, regulär ist. Daher ist $|X^{PGL(2,q)}| \geqslant q(q - 1)$, so daß $|PGL(2,q)_X| \leq q + 1$ ist. Wegen $U \subseteq PGL(2,q)_X$ ist daher $U = PGL(2,q)_X$. Nun ist $PSL(2,q)_X \subseteq U$. Ist q gerade, so ist $PSL(2,q) = PGL(2,q)$. Ist q ungerade, so enthält $PSL(2,q)$ keine zyklische Untergruppe der Ordnung $q + 1$. Weil

Heinz Lüneburg

$PSL(2,q)_X$ somit eine echte Untergruppe von U ist, ist ihre Ordnung höchstens gleich $\frac{1}{2}(q + 1)$. Daher ist $|X^{PSL(2,q)}| \geqslant q(q - 1)$, so daß in jedem Falle auch $PSL(2,q)$ auf der Menge der Punkte von π, die nicht in $V_0(2,q)$ liegen, transitiv operiert. Diese Bemerkung benutzen wir beim Beweise von

4.6. Hilfssatz. Sind π und π' zwei desarguessche Kongruenzpartitionen von $V(4,q)$ und ist ϱ eine Regelschar, die in π und in π' enthalten ist, ist ferner $|\pi \cap \pi'| \geqslant q + 1$, so ist $\pi = \pi'$.

Beweis. Es sei π' die zu π konjugierte Regelschar. Ferner sei G die Gruppe, die π' elementweise festläßt. Dann ist $G \cong GL(2,q)$. Hilfssatz 4.1 liefert die Existenz zweier Untergruppen H und H_1 von G, die beide zur $SL(2,q)$ isomorph sind und die π bzw. π' invariant lassen. Weil $GL(2,q)$ nur eine zur $SL(2,q)$ isomorphe Untergruppe enthält, folgt $H = H_1$. Ist Nun $X \in \pi \cap \pi'$ und $X \notin \varrho$, so folgt nach unserer Vorbemerkung $|X^H| = q(q - 1)$. Folglich ist $\pi = \varrho \cup X^H = \pi'$, q. e. d.

4.7. Satz (Prohaska). Es sei p eine Primzahl und A eine Translationsebene der Ordnung p^2. Ist Σ eine von Elationen erzeugte Kollineationsgruppe von A, die zur $SL(2,q)$ isomorph ist, so ist A desarguessch.

Beweis. Weil A eine Translationsebene ist, können wir o. B. d. A. annehmen, daß Σ einen Fixpunkt hat. P sei dieser Fixpunkt. Ferner folgt, weil A eine Translationsebene ist, daß alle Elationen von A, die von 1 verschieden sind, die Ordnung p haben. Ist nun Π eine p-Sylowgruppe von Σ, so hat Π eine Fixgerade, da $p^2 + 1$ die Anzahl der Geraden durch P ist. Weil Σ von Elationen erzeugt wird und nicht triviale Elationen die Ordnung p haben, folgt, daß Π nur aus Elationen besteht, da alle Elemente der Ordnung p in Σ konjugiert sind. Hätte nun eine weitere p-Sylowgruppe die gleiche Fixgerade wie Π, so bestünde Σ nur aus Elationen, da Σ von irgendzwei seiner p-Sylowgruppen erzeugt wird. Dies kann aber nicht sein, da Σ dann eine p-Gruppe wäre. Also haben verschiedene p-Sylowgruppen verschiedene Fixgeraden. Weil die Anzahl der p-Sylowgruppen von Σ gleich p + 1 ist, erfüllen Σ, der Punkt P und die Menge $\wp$ der Schnittpunkte der Fixgeraden der p-Sylowgruppen mit der uneigentlichen Geraden die Voraussetzungen von Hilfssatz 4.1. Ist daher Q ein Punkt mit $P \neq Q \notin \wp$, sowie $PQ \cap g_\infty \in \wp$, so ist $\{P\} \cup Q^\Sigma \cup \wp$ die Punktmenge einer Baerunterebene von A. Von diesen Baerunterebenen gibt es p + 1 Stück. $T_1, \ldots, T_{p+1}$ seien die Stabilisatoren dieser Unterebenen in der Translationsgruppe T von A. Diese Stabilisatoren haben alle die Ordnung p^2. Ferner hat T die Ordnung p^4, so daß T ein Vektorraum vom Rang 4 über GF(p) ist. Die T_i sind dann Unterräume vom Rang 2 von T. Ferner folgt, daß die T_i gerade die sämtlichen Transversalen der $T(X, g_\infty)$ mit $X \in \wp$ sind. Folglich bilden die $T(X, g_\infty)$ eine Regelschar. Diese werde mit ς bezeichnet. Ferner sei $\pi = \{T(Y, g_\infty) | Y \, I \, g_\infty\}$. Es sei nun $U \in \pi \setminus \wp$ und es seien $\pi_1, \ldots, \pi_{\frac{1}{2}q(q-1)}$ die sämtlichen desarguesschen Regelscharen, die ς enthalten. Nach 4.6 ist dann $\pi_i \cap \pi_j = \varsigma$ für $i \neq j$. Nach 4.2 gibt es daher ein i mit $U \in \pi_i$. Nach der Bemerkung vor 4.6 ist daher $\pi_i \setminus \varsigma = U^\Sigma \subseteq \pi \setminus \varsigma$ und folglich $\pi = \pi_i$, q. e. d.

Heinz Lüneburg

5. Ableitbare Netze. $N = (\mathcal{R}, \mathcal{N}, I)$ und $N' = (\mathcal{R}, \mathcal{N}', I')$ seien Netze auf der selben Punktmenge $\mathcal{R}$. Das Netz N heißt ersetzbar durch das Netz N', falls zwei verschiedene Punkte aus $\mathcal{R}$ genau dann eine Verbindungsgerade in $\mathcal{N}$ haben, wenn sie eine Verbindungsgerade in $\mathcal{N}'$ haben.

Ist das Netz $N = (\mathcal{R}, \mathcal{N}, I^0)$ Unterstruktur der affinen Ebene $A = (\mathcal{R}, \mathcal{G}, I)$ und ist N ersetzbar durch das Netz $N' = (\mathcal{R}, \mathcal{N}', I')$, so ist auch $A' = (\mathcal{R}, \mathcal{G}', I^*)$ mit $\mathcal{G}' = (\mathcal{G} \setminus \mathcal{N}) \cup \mathcal{N}'$ und $I^* = (I \setminus I^0) \cup I'$ eine affine Ebene. In diesem Falle schreiben wir $A' = A(N/N')$.

Ist $N = (\mathcal{R}, \mathcal{N}, I)$ durch $N' = (\mathcal{R}, \mathcal{N}', I')$ ersetzbar und ist $g \in \mathcal{N}'$, so sei

$$N(g) = (I'(g), \{x | x \in \mathcal{N}, I(x) \cap I'(g) \neq \emptyset\}, I' \cap \ldots).$$

Dabei stehen die Pünktchen für das cartesische Produkt der Punkt- mit der Geradenmenge von N(g). N heißt ableitbar, falls N(g) für alle $g \in \mathcal{N}'$ eine affine Ebene ist. Einfaches Zählen zeigt, daß dann auch N'(h), was analog zu N(g) definiert wird, für alle Geraden h von N eine affine Ebene ist. Ist N in A eingebettet und ist N durch N' ableitbar, so heißt auch A ableitbar und A(N/N') heißt die abgeleitete Ebene. Ist n die Ordnung von A, so ist n auch die Ordnung von N und N'. Hieraus folgt, daß n die Anzahl der Punkte in N(g) ist. Folglich ist $n = m^2$ und N(g) ist eine Baerunterebene von A. Aus Hilfssatz 3.2 folgt, wie Ostrom bemerkte, daß A(N/N') durch A und N bis auf Isomorphie eindeutig bestimmt ist. Um zu sehen, daß sich 3.2 anwenden läßt, muß man sich nur überlegen, daß

Heinz Lüneburg

N(g) und N(h) für alle g,h ∈ 𝔐 ' die gleichen uneigentlichen Punkte
besitzen: Um dies zu zeigen, sei zunächst $I'(g) \cap I'(h) = \emptyset$. Weil
$I'(g)$ eine affine Ebene ist, gibt es $m^2 + m$ Geraden von N, die Geraden von N(g) sind. Haben zwei dieser Geraden einen Schnittpunkt,
so ist dies ein Punkt von N(g). Also tragen diese Geraden insge-
samt $(m^2 + m)(m^2 - m) = m^4 - m^2 = n^2 - m^2$ Punkte von N, die nicht
zu g gehören. Dies sind aber alle Punkte, die nicht auf g liegen.
Es gibt also eine Gerade a von N, die sowohl Punkte von g als auch
Punkte von h trägt. Weil N durch N' ersetzbar ist, gibt es daher
eine Gerade l von N', die mit g und auch mit h einen Punkt gemeinsam hat. Wir können daher o. B. d. A. annehmen, daß g und h einen
Punkt gemeinsam haben. Es sei P ein solcher Punkt. Die m + 1 Geraden von N durch P sind dann sowohl Geraden von N(g) als auch von
N(h), so daß die m + 1 uneigentlichen Punkte von N(g) mit denen
von N(h) übereinstimmen.

Beispiel. Es sei ϱ eine Regelschar in $V = V(4,q)$ und $\overline{ϱ}$ sei die
zu ϱ konjugierte Regelschar. Ist dann

$$N = (V, \{ X + v \mid X \in ϱ , v \in V \}, \in)$$

und

$$N' = (V, \{ Y + v \mid Y \in \overline{ϱ} , v \in V \}, \in),$$

so ist N durch N' ableitbar. Weil ϱ , wie wir wissen, in einer desarguesschen Kongruenzpartition enthalten ist, folgt, daß die desarguessche Ebene A der Ordnung q^2 ableitbar ist. A(N/N') ist die Hall-
ebene der Ordnung q^2.

Heinz Lüneburg

Wie wir schon bemerkten, ist die Ordnung n eines durch N' ableit-
baren Netzes N Quadrat einer natürlichen Zahl m. Ferner ist r = m + 1.
Weiterhin gilt

j) Ist g eine Gerade von N' und P $\not I$ 'g, so gibt es genau eine Gerade
h von N(g) mit P I h.

Dies haben wir ebenfalls schon bewiesen, als wir zeigten, daß 3.2
beim Beweise der Eindeutigkeit von A(N/N') anwendbar ist.

ij) g und h seien Geraden von N', die keinen Punkt gemeinsam haben.
Es gibt dann genau eine Parallelenschar $\wp$ von N mit den Eigenschaften:
(a) Ist a eine Gerade von N, die Gerade von N(g) und N(h) ist, so
ist a $\in \wp$.
(b) Ist a $\in \wp$, so ist a genau dann eine Gerade von N(g), wenn a
Gerade von N(h) ist.

Beweis. Sind a,b zwei verschiedene Geraden, die beide Geraden von
N(g) und N(h) sind, so ist a $\cap$ b = $\emptyset$, da ein Schnittpunkt von a mit
b Punkt von N(g) und Punkt von N(h) wäre. Also gibt es höchstens ein
solches $\wp$. Es sei nun P I' g. Nach j) gibt es eine Gerade a durch
P, welche Gerade von N(h) ist. Ist $\wp$ die durch a bestimmte Parallelenschar, so hat $\wp$ offenbar alle gewünschten Eigenschaften.

Wir führen noch eine weitere Bezeichnung ein. Ist g eine Gerade von
N und h eine Gerade von N', so sei

Heinz Lüneburg

$$N(h,g) = (I'(h), \{x \,|\, x \in \mathcal{N} \,,\, I(x) \cap I(g) \neq \emptyset \neq I(x) \cap I'(h)\}, I \cap \ldots).$$

Entsprechend werde $N'(g,h)$ definiert. Offenbar ist $N(h,g)$ die Inzidenzstruktur, die aus $N(h)$ entsteht, in dem man die zu g parallelen
Geraden von $N(h)$ aus $N(h)$ entfernt. Dies beweist den ersten Teil des
folgenden Hilfssatzes

iij) $N(h,g)$ ist ein Netz der Ordnung m und des Gerades m. Ist der
Durchschnitt von $I'(h)$ mit $I(g)$ leer, so ist $N(h,g)$ dual zu $N'(g,h)$.

Beweis. Es bleibt nur zu zeigen, daß $N(h,g)$ und $N'(g,h)$ dual zuein-
ander sind. Es sei $P \in I'(h)$. Nach Voraussetzung ist dann $P \notin I(g)$.
Nach j) gibt es daher eine eindeutig bestimmte Gerade P^δ von N'
mit $P \ I' \ P^\delta$, die Gerade von $N'(g)$ ist. Wegen $P \ I' \ P^\delta$ und weil
$I'(P^\delta) \cap I(g) \neq \emptyset$ ist, ist P^δ eine Gerade von $N'(g,h)$. Es sei y
eine von g verschiedene Gerade von N, die g schneidet. Es gibt dann
einen und nur einen Punkt y^δ mit $y^\delta \ I \ y$ und $y^\delta \ I \ g$. Ist nun $P \ I \ y$,
so gibt es, weil ja auch y^δ auf y liegt, eine Gerade $x \in \mathcal{N}'$ mit
$P, y^\delta \ I' \ x$, denn N wird ja durch N' ersetzt. Nun hat x mit g den Punkt
y^δ gemein. Folglich ist x eine Gerade von $N'(g)$ und damit nach j) die
einzige Gerade von $N'(g)$, die durch P geht. Also ist $x = P^\delta$. Aus
$P \ I \ y$ folgt also $y^\delta \ I' \ P^\delta$. Weil alles in N und N' symmetrisch ist,
folgt,daß die Abbildung δ eine Inverse besitzt, die ebenfalls inzi-
denztreu ist, so daß δ in der Tat eine Dualität von $N(h,g)$ auf $N'(g,h)$
ist, q. e. d.

Im folgenden bezeichnen wir die soeben konstruierte Dualität mit $\delta(h,g)$.

Heinz Lüneburg

Offenbar ist $\delta(g,h)$ zu $\delta(h,g)$ invers.

Im folgenden benötigen wir drei Sätze, die wir hier ohne Beweis angeben.

5.1. Satz (A. Wagner) Es sei A eine endliche affine Ebene und Γ sei eine Kollineationsgruppe von A. Operiert Γ auf der Menge der Geraden von A transitiv, so ist A eine Translationsebene und Γ enthält die Translationsgruppe von A.

Für einen Beweis siehe WAGNER [9] oder auch LÜNEBURG [6].

5.2. Satz (Skornjakov & San Soucie). Ist Π eine projektive Ebene und g eine Gerade von Π, ist ferner $(\Pi^d)^P$ eine Translationsebene für alle P I g, so ist Π^h Translationsebene für alle Geraden h von Π, dh., Π ist eine Moufangebene.

Einen Beweis dieses Satzes, der den Fall Charakteristik 2 und Charakteristik ungleich 2 gleichzeitig erledigt, findet der Leser in LÜNEBURG [6].

5.3. Satz (Artin & Zorn). Ist Π eine endliche Moufangebene, so ist Π desarguessch.

Ein Beweis dieses Satzes findet sich in PICKERT [7] oder auch in LÜNEBURG [6].

Heinz Lüneburg

5.4. Satz (Prohaska). Ist h eine Gerade von N', so ist die affine
Ebene N(h) desarguessch.

Beweis. Wir können o. B. d. A. annehmen, daß $m > 2$ ist, denn es
gibt ja nur eine affine Ebene der Ordnung 2. Dann ist also $m + 1 \geqslant 4$,
so daß N mindestens vier Parallelenscharen hat.

x und y seien zwei Geraden von N(h). Es gibt dann eine Gerade u von
N(h), die weder zu x noch zu y parallel ist. Wählt man einen Punkt
auf u, der nicht zu N(h) gehört, so findet man eine zu h parallele
Gerade v in N', die h nicht trifft. Es sei $\wp$ die durch u bestimmte Parallelenschar von N. Dieses $\wp$ hat dann die Eigenschaften von
ij). Wegen $m \geqslant 3$ gibt es schließlich eine Gerade z von N(v), die
weder zu x noch zu y noch zu u parallel ist. Schließlich seien
$a,b \in \wp$ und X I a sowie Y I b. Es folgt $I(a) \cap I'(h) = \emptyset$. Andern-
falls wäre a eine Gerade von N(h) und daher $X \in I'(h)$. Dann wäre
aber z eine Gerade von N(h), da X auf z liegt. Weil z dann sowohl
Gerade von N(h) als auch Gerade von N(v) wäre, wäre $z \in \wp$: ein
Widerspruch. Entsprechend folgt $I(a) \cap I'(v) = \emptyset$. Somit sind
$\delta(h,a)$ und $\delta(a,v)$ definiert. Nun ist N'(a,h) die Inzidenzstruktur,
die aus N'(a) durch Entfernen der zu h parallelen Geraden entsteht,
und N'(a,v) die Inzidenzstruktur, die aus N'(a) durch Entfernen der
zu v parallelen Geraden entsteht. Weil h zu v parallel ist, folgt
N'(a,h) = N'(a,v). Somit ist das Produkt $\sigma = \delta(h,a)\,\delta(a,v)$ definiert und ist ein Isomorphismus von N(h,a) auf N(v,a), welcher offenbar x auf z abbildet. Vertauscht man nun die Rollen von v und h und
ersetzt man a durch b, so erhält man einen Isomorphismus τ von

Heinz Lüneburg

N(v,b) auf N(h,b) mit z^τ = y. Weil a zu b parallel ist, folgt
N(v,b) = N(v,a) und N(h,b) = N(h,a). Somit ist $\sigma\tau$ eine Kollinea-
tion von N(h,a), die x auf y abbildet. Nun ist N(h,a) bis auf eine
Parallelenschar dasselbe wie N(h) und man sieht leicht, daß sich
$\sigma\tau$ zu einer Kollineation von N(h) fortsetzen läßt. Weil die
Kollineationsgruppe von N(h) also auf der Menge der Geraden von
N(h) transitiv operiert, ist N(h) nach 5.1 eine Translationsebene.

Ebenso gilt natürlich, daß N'(g) für alle Geraden g von N eine Translationsebene ist. Es sei 1 irgendeine Gerade von N(h). Ferner sei
g eine zu 1 parallele Gerade von N, die mit h keinen Punkt gemeinsam hat. Dann ist N'(g) eine Translationsebene und N'(g,h) ist zu
N(h,g) dual. Ist nun π der projektive Abschluß von N(h) und P der
Schnittpunkt von 1 mit der uneigentlichen Geraden, so folgt aus der
dualen Isomorphie von N'(g,h) mit N(h,g), daß $(\pi^d)^P$ eine Translationsebene ist. Aus 5.2 und 5.3 folgt nun, daß N(h) desarguessch
ist, q. e. d.

6. Eine Charakterisierung der Hallebenen. Wir beginnen wieder mit
mehreren Hilfssätzen.

6.1. Hilfssatz. A sei eine desarguessche affine Ebene der Ordnung
q = p^r. Ist π eine p-Untergruppe der Kollineationsgruppe von A,
hat π einen Fixpunkt und ist $|\pi| \geqslant$ q, so enthält π eine Elation,
deren Achse eine Gerade von A ist. (Solche Elationen heißen Scherungen.)

Heinz Lüneburg

Beweis. P sei der Fixpunkt von Π und Γ sei die Gruppe der Kollineationen von A, die P zum Fixpunkt haben. Dann ist $\Gamma \cong \Gamma L(2,q)$. Folg-
lich ist $|\Gamma| = rq(q^2 - 1)(q - 1)$. Weil q + 1 die Anzahl der Geraden
durch P ist, hat Π eine Fixgerade g, die mit P inzidiert. Es sei T
die Gruppe aller Scherungen mit der Achse g. Dann ist ΠT eine
p-Gruppe, da T ein Normalteiler von Γ_g ist. Ist p^s die höchste Po-
tenz von p, die in r aufgeht, so ist $|\Pi T|$ ein Teiler von p^{s+r}. Nun
ist $p^s \leq r < p^r$, so daß $|\Pi T| < q^2$ ist. Also ist $\Pi \cap T \neq \{1\}$, q.e.d.

Im folgenden sei stets A = $(\mathcal{R}, \mathcal{G}, I)$ eine affine Ebene und N = $(\mathcal{R},$
$\mathcal{M}, I_o)$ ein in A eingebettetes, ableitbares Netz. Ferner sei N durch
N' = $(\mathcal{R}, \mathcal{M}', I_o')$ abgeleitet und A' = A(N/N') = $(\mathcal{R}, \mathcal{G}', I')$. Ist n
die gemeinsame Ordnung von A und A', so ist, wie wir nun wissen,
n = q^2 und q = p^r mit einer Primzahl p. Ist q = 2, so haben A und
A' die Ordnung 4 und sind folglich beide desarguessch. Wir nehmen
daher im folgenden an, daß q > 2 ist.

6.2. Hilfssatz (Prohaska). η sei eine Kollineation von A mit dem
Fixpunkt O. Läßt η jede Gerade von I(0) \ $I_o(0)$ fest, so ist η
eine Homologie von A oder von A'.

Beweis. Weil die Geradenmenge von N Vereinigung von vollen Paralle-
lenklassen von A ist und weil N durch N' abgeleitet wird, ist η sowohl Kollineation von A als auch von A'. Als Kollineation von A läßt
η das Netz N invariant und als Kollineation von A' das Netz N'.

Läßt η einen von O verschiedenen Punkt von A fest, so läßt η in der zu A gehörenden projektiven Ebene ein Viereck und damit eine Unterebene punkt- und geradenweise fest. Weil die $q(q - 1)$ Geraden von $I(0) \smallsetminus I_0(0)$ zu dieser Unterebene gehören, ist die Ordnung dieser Unterebene mindestens gleich $q(q - 1) - 1$. Weil $q > 2$ ist, folgt $q(q - 1) - 1 > q$, so daß diese Unterebene nach 3.1 die ganze Ebene ist. In diesem Falle ist also $\eta = 1$, so daß nichts mehr zu beweisen ist.

Wir nehmen nun an, daß η keine Homologie von A ist. Dann gibt es ein $g \in I_0(0)$ mit $g^\eta \neq g$. Insbesondere ist dann $\eta \neq 1$, so daß 0 der einzige Fixpunkt von η ist. Es sei $h \in I_0'(0)$ und $h^\eta \neq h$. Wegen $0 \in I_0(g) \cap I_0'(h)$ ist g eine Gerade von $N(h)$. Es gibt somit einen Punkt $P \in I_0(g) \cap I_0'(h)$ mit $P \neq 0$. Wie wir schon bemerkten, ist $P \neq P^\eta$. Nach 3.2 haben $N(h)$ und $N(h^\eta)$ nur einen affinen Punkt gemeinsam, nämlich 0. Die Geraden von N durch 0 sind daher die einzigen Geraden, die $N(h)$ und $N(h^\eta)$ gemeinsam haben. Wäre nun PP^η eine Gerade von N, so wäre PP^η eine Gerade von $N(h)$ und von $N(h^\eta)$, weil P auf h und P^η auf h^η liegt. Daher wäre $0 \text{ I } PP^\eta$ und folglich $g = OP = OP^\eta$. Hieraus folgte $g^\eta = g$: ein Widerspruch. Also ist PP^η keine Gerade von N. Hieraus folgt wiederum, daß der uneigentliche Punkt auf PP^η ein Fixpunkt von η ist. Dies impliziert seinerseits $(PP^\eta)^\eta = PP^\eta$. Dies hat wiederum zur Folge, daß η einen von 0 verschiedenen Fixpunkt in A hat. Dieser Widerspruch zeigt, daß $h^\eta = h$ ist für alle $h \in I_0'(0)$. Nun ist $I(0) \smallsetminus I_0(0) = I'(0) \smallsetminus I_0'(0)$, so daß η alle Geraden von A', die durch 0 gehen einzeln invariant läßt, q. e. d.

Heinz Lüneburg

6.3. Hilfssatz. A sei eine desarguessche affine Ebene der Ordnung q. Ferner sei $\triangle$ eine Kollineationsgruppe von A, die den Punkt O festläßt. Wird $\triangle$ von Elationen erzeugt und enthält $\triangle$ zu jeder Geraden durch O eine von 1 verschiedene Elation mit dieser Geraden als Achse, so ist $\triangle \cong SL(2,q)$ oder aber q ist gerade und $\triangle$ ist eine Diedergruppe der Ordnung 2(q + 1).

Beweis. Weil $\triangle$ von Elationen erzeugt wird, deren Achsen alle durch O gehen, und alle solche Elationen in einer zur SL(2,q) isomorphen Gruppe liegen, folgt, daß $\triangle$ zu einer Untergruppe der SL(2,q) isomorph ist. Ferner folgt, daß q + 1 die Anzahl der p-Sylowgruppen von $\triangle$ ist, wenn $q = p^r$ und p eine Primzahl ist. Nach DICKSON [3] folgt nun, daß entweder $\triangle \cong SL(2,q)$ oder daß q gerade und $\triangle$ eine Diedergruppe der Ordnung 2(q + 1) ist.

6.4. Hilfssatz. Es sei $\mathcal{L}$ eine Menge mit $|\mathcal{L}| = q(q - 1)$. Ferner sei q Potenz einer Primzahl p und $\sum$ sei eine zur PSL(2,q) isomorphe Permutationsgruppe von $\mathcal{L}$. Haben alle Bahnen von $\sum$ die gleiche Länge und hat kein Element der Ordnung p ein Fixelement in $\mathcal{L}$, so gibt es eine Zerlegung π von $\mathcal{L}$ in 2-Teilmengen, die von $\sum$ invariant gelassen wird, so daß der Stabilisator eines $X \in \pi$ eine maximale Diedergruppe von $\sum$ ist, es sei denn, es ist

a) $q \in \{5,7,11,23\}$ und $\sum_b \cong A_4$ für $b \in \mathcal{L}$.

b) $q \in \{7,23,47\}$ und $\sum_b \cong S_4$ für $b \in \mathcal{L}$.

c) $q \in \{11,19,29,59\}$ und $\sum_b \cong A_5$ für $b \in \mathcal{L}$.

Dies ist eine Übungsaufgabe, die mit Hilfe der Untergruppenliste der

Heinz Lüneburg

PSL$(2,q)$ rasch und einfach zu lösen ist (s. DICKSON, loc. cit.).

6.5. Satz (Prohaska). A, N, N' und A' haben die gleichen Bedeutungen
wie bisher. Hat A eine Kollineationsgruppe, die N invariant läßt und
die als Gruppe vom Rang 3 auf den Punkten von A operiert, so ist eine
der beiden Ebenen A und A' desarguessch und die andere ist eine Hall-
ebene.

Beweis. Offenbar gelten die Voraussetzungen für A,N genau dann, wenn
sie für A',N' gelten. Nach KALLAHER [4] und LIEBLER [5] sind daher A und A' Translationsebenen.

Es sei nun 0 ein Punkt von A. Nach Voraussetzung gibt es dann eine
Kollineationsgruppe $\sum$ von A, die 0 festläßt und die außer $\{0\}$ noch
genau zwei Punktbahnen hat. Es sei $\mathcal{a} = I_0(0)$, $\mathcal{b} = I(0) \setminus I_0(0)$
und $\mathcal{a}' = I_0'(0)$. Offenbar sind $\mathcal{a}$, $\mathcal{a}'$ und $\mathcal{b}$ unter $\overline{}$ invariant.
Überdies folgt, wie man sich leicht überlegt, daß $\mathcal{a}$, $\mathcal{a}'$ und $\mathcal{b}$
Geradenbahnen von $\sum$ sind. Ferner sind

$$\mathcal{u} = \{ P | P \in \mathcal{R},\ 0 \neq P\ I\ a,\ a \in \mathcal{a} \}$$

und

$$\mathcal{l} = \{ P | P \in \mathcal{R},\ 0 \neq P\ I\ b,\ b \in \mathcal{b} \}$$

die beiden von $\{0\}$ verschiedenen Punktbahnen von $\sum$. Es ist
$|\mathcal{a}| = q + 1 = |\mathcal{a}'|$ und $|\mathcal{b}| = q(q - 1)$ und daher
$|\mathcal{u}| = (q + 1)(q^2 - 1)$ und $|\mathcal{l}| = q(q - 1)(q^2 - 1)$. Hieraus folgt,
daß das kleinste gemeinsame Vielfache von $(q + 1)(q^2 - 1)$ und
$q(q - 1)(q^2 - 1)$ ein Teiler von $|\sum|$ ist. Ist $d = (2, q - 1)$, so ist

Heinz Lüneburg

dieses kleinste gemeinsame Vielfache gleich $d^{-1}q(q^2 - 1)^2$.

Als nächstes zeigen wir, daß Σ eine Scherung von A oder von A' enthält. Angenommen, dies ist nicht der Fall. Dann sei Π eine p-Sylowgruppe von Σ. Weil q ein Teiler von $|\Sigma|$ ist, ist q ein Teiler von $|\Pi|$. Ferner gibt es ein $a \in \mathcal{O}$ mit $a^{\Pi} = a$, da ja $|\mathcal{O}| = q + 1$ ist. Ebenso gibt es ein $b \in \mathcal{O}'$ mit $b^{\Pi} = b$. Weil Π keine Scherung enthält, operiert Π treu auf den Punkten von N(b). Weil N(b) nach 5.4 desarguessch ist, enthält Π nach 6.1 ein π, welches als nicht triviale Scherung auf N(b) wirkt. Offenbar ist dann a die Achse dieser Scherung. Folglich ist jeder Punkt von $I(a) \cap I'(b)$ ein Fixpunkt von π, falls π das fragliche Element von Π ist. Angenommen Σ_a operiere nicht treu auf I(a). Dann enthält Σ_a eine von 1 verschiedene Perspektivität mit der Achse a. Diese Perspektivität läßt eine Parallelenschar $\mathcal{P}$ von A geradenweise fest. Nach Annahme ist $a \notin \mathcal{P}$. Weil Σ_a keine Scherung mit der Achse a enthält, müssen nach ANDRÉ $[2]$ alle Kollineationen aus Σ_a auch $\mathcal{P}$ festlassen. Umgekehrt müssen aus dem gleichen Grunde, da ja $\mathcal{O}$ eine Bahn von Σ ist auch alle Kollineationen aus $\Sigma_{\mathcal{P}}$ die Gerade a festlassen. Also ist $\Sigma_{\mathcal{P}} = \Sigma_a$. Insbesondere folgt $\mathcal{P}^{\pi} = \mathcal{P}$, so daß π auf N(b) trivial operiert. Dieser Widerspruch zeigt, daß Σ_a auf I(a) treu operiert. Ebenso folgt, daß Σ_b auf I'(b) treu operiert.

Es sei $\Delta = \langle \pi^{\alpha} \mid \pi \in \Pi, \alpha \in \Sigma_b, \pi$ induziert eine von 1 verschiedene Scherung in N(b)$\rangle$. Mit Hilfe von 5.4 und 6.3 ist entweder Δ zur SL(2,q) isomorph oder q ist gerade und Δ ist eine Diedergruppe der Ordnung $2(q + 1)$. Hierbei haben wir benutzt, daß $\triangle$ treu induziert wird, da ja $\sum_b$ auf $I'(b)$ treu operiert. Nun ist

$$|\textstyle\sum| = |b^{\sum}||\textstyle\sum_b| = (q + 1)|\textstyle\sum_b|.$$

Weil $d^{-1}q(q^2 - 1)^2$ ein Teiler von $|\sum|$ ist, ist daher $d^{-1}q(q-1)(q^2-1)$ ein Teiler von $|\sum_b|$. Wäre nun $\triangle$ eine Diedergruppe der Ordnung $2(q + 1)$, so wäre $|\sum_b|$ ein Teiler von $2r(q + 1)$, wobei r durch $q = 2^r$ bestimmt ist. Dann wäre $q(q - 1)^2$ ein Teiler von $2r$, woraus wegen $r < 2^r = q$ die Ungleichung $q(q - 1)^2 < 2q$ folgte. Dies implizierte $q = 2$, was wir ausgeschlossen haben. Also ist $\triangle \cong SL(2,q)$. Es sei nun Ψ eine p-Sylowgruppe von $\triangle$. Dann ist $|\Psi| = q$. Ferner gibt es eine Gerade von $\mathcal{U}$, die wir wieder a nennen, die von Ψ festgelassen wird. Ψ induziert dann in $N(b)$ eine Gruppe von Scherungen mit der Achse a. Weil wir angenommen haben, daß $\sum$ keine Scherungen von A enthält, induziert Ψ auch eine Gruppe von Scherungen in $N'(a)$, die dann die Achse b haben. Es folgt, daß $\triangle$ auf $\mathcal{U}' \smallsetminus \{b\}$ transitiv ist. Also enthält $\triangle$ eine Untergruppe vom Index q, so daß q nach DICKSON, loc. cit. eine der Primzahlen 3,5, 7 oder 11 ist. Weil $q = p$ eine Primzahl ist, ist Ψ auch eine p-Sylowgruppe von $\sum_b$ und damit wegen $|\sum : \sum_b| = q + 1$ auch von $\sum$. Somit operiert Ψ auf $\mathcal{l}$ wegen $|\mathcal{l}| = q(q - 1)(q^2 - 1)$ regulär. Nun operiert $\pi \in \Psi$ auf $N'(a)$ als Scherung mit der Achse b. Folglich läßt π jede zu b parallele Gerade c von $N'(a)$ fest. Weil π auch alle zu a parallelen Geraden von $N(b)$ festläßt, folgt mit i,j), daß π in $N(c)$ eine nicht triviale Translation oder Scherung induziert. Im ersten Falle läßt π alle Parallelenscharen von N fest, was zur Folge hat, daß π auf $N(b)$ trivial

Heinz Lüneburg

operiert; ein Widerspruch. Im zweiten Fall hat π Fixpunkte P und Q
auf b bzw. c. Nach ij) ist PQ $\parallel$ a und daher PQ = a. Hieraus folgt,
daß π in N'(a) mehr als q Fixpunkte hat, woraus wiederum π = 1
folgt. Dieser letzte Widerspruch zeigt, daß $\sum$ doch eine Scherung
von A oder von A' enthält.

Wir können aus Symmetriegründen annehmen, daß $\sum$ eine von 1 verschiedene Scherung von A enthält. Es sei $\triangle$ die von den Scherungen aus
$\sum$ erzeugte Gruppe. Dann folgt, daß $\triangle$ jede Gerade aus $\mathcal{O}'$ festläßt. Ferner ist $\triangle$ treu auf I'(b) für alle b $\in \mathcal{O}'$. Es folgt wieder $\triangle \cong$ SL(2,q) oder q ist gerade und $\triangle$ ist eine Diedergruppe der
Ordnung 2(q + 1). Weil $\triangle$ von den Scherungen aus $\sum$ erzeugt wird,
ist $\triangle$ normal in $\sum$. Hieraus folgt, daß die Bahnen von $\triangle$ xxxxxxx
in $\mathcal{L}$ alle die gleiche Länge haben, da $\mathcal{L}$ ja eine Bahn von $\sum$ ist.
Ist $\triangle$ die Diedergruppe der Ordnung 2(q + 1) und ist t die Länge
der Bahnen von $\triangle$ in $\mathcal{L}$, so ist t ein Teiler von $(|\triangle| , |\mathcal{L}|)$ =
= $(2(q + 1), q(q - 1))$ = 2, so daß $\mathcal{L}$ unter $\triangle$ in lauter Bahnen der
Länge 2 zerfällt, da die Elemente der Ordnung 2 ja Scherungen sind
und in $\mathcal{L}$ folglich keine Fixgeraden haben. Der zyklische Normalteiler Z der Ordnung q + 1 von $\triangle$ läßt daher alle Geraden von $\mathcal{L}$
fest. Weil Z auf $\mathcal{O}$ transitiv operiert, folgt nach 6.2, daß Z
aus Homologien von A' besteht. Der Kern der Translationsebene A'
enthält also eine zyklische Untergruppe der Ordnung q + 1, woraus
folgt, daß dieser Kern zu $GF(q^2)$ isomorph ist. In diesem Falle ist
also A' desarguessch.

Es sei nun $\triangle \cong$ SL(2,q). Ist q = p, so folgt aus 4.7, daß A desarguessch

Heinz Lüneburg

ist. Es sei also $q = p^r > p$. Die von Δ auf $\mathscr{L}$ induzierte Gruppe
ist isomorph zur PSL(2,q). Weil Σ auf $\mathscr{L}$ transitiv operiert, folgt
wieder, daß alle Bahnen von Δ in $\mathscr{L}$ die gleiche Länge haben. Nach
6.4 gibt es wegen $q > p$ eine Zerlegung π von $\mathscr{L}$ in 2-Teilmengen,
die von Δ invariant gelassen wird. Es sei Z der Zentralisator von
Δ in Σ. Dann ist Σ / Z isomorph zu einer Untergruppe von
$\mathrm{P\Gamma L}(2,q)$, da $\mathrm{P\Gamma L}(2,q)$ die Automorphismengruppe von SL(2,q) ist.
Hieraus folgt, daß $|\Sigma / Z|$ ein Teiler von $rq(q^2 - 1)$ ist. Also
ist $|\Sigma / Z| t = rq(q^2 - 1)$. Ferner ist $|\Sigma| = ud^{-1}q(q^2 - 1)^2$, so
daß $q^2 - 1$ ein Teiler von $rd|\Sigma|$ ist. Nun operiert Z auf π trivial,
da Z ja insbesondere alle maximalen Diedergruppen von Δ zentrali-
siert. Hieraus folgt, daß die Untergruppe Λ von Z, die alle Geraden von $\mathscr{L}$ festläßt, höchstens den Index 2 in Z hat. Nach 6.1 ist
jedes Element von Λ entweder eine Homologie von A oder eine von A'.
Weil Λ nicht Vereinigung von zwei echten Untergruppen sei kann,
folgt, daß Λ entweder nur aus Homologien von A oder nur aus Homologien von A' besteht. (Es ist nicht verboten, daß Λ Elemente enthält, die Homologien von A und von A' sind.) Nun ist $dr < q - 1$
und folglich $q + 1 < |Z|$, woraus $\frac{1}{2}(q + 1) < |\Lambda|$ folgt. Dies impli-
ziert wieder, daß der Kern von A oder von A' gleich $GF(q^2)$ ist, q.e.d.

Literatur.

[1] André, J., Über nicht desarguessche Ebenen mit transitiver Translationsgruppe. Math. Z. 60, (1954), 156-186.

[2] André, j., Über Perspektivitäten in endlichen projektiven Ebenen.
Arch. Math. 6(1954), 29-32.

[3] Dickson, L. E., Linear groups. Neudruck: New York 1958.

[4] Kallaher, M. J., On finite affine planes of rank 3. J. of Algebra

Heinz Lüneburg

13 (1969), 544-553.

[5] Liebler, R. A., Finite affine planes of rank three are translation planes. Math. Z. 116 (1970), 89 - 93.

[6] Lüneburg, H., Lectures on projective planes. University of Illinois at Chicago Circle. Chicago 1969.

[7] Pickert, G., Projektive Ebenen. Berlin-Göttingen-Heidelberg 1955.

[8] Prohaska, O., Endliche ableitbare affine Ebenen. Geometriae Dedicata 1 (1972).

[9] Wagner, A., On finite affine line transitive planes. Math. Z. 87 (1965), 1-11.

CENTRO INTERNAZIONALE MATEMATICO ESTIVO
(C. I. M. E.)

J. A. THAS

4-GONAL CONFIGURATIONS

Corso tenuto a Bressanone dal 18 al 27 Giugno 1972

4-GONAL CONFIGURATIONS

J.A.THAS

1. All known 4-gonal configurations

1.1. A finite 4-gonal configuration [4] is an incidence structure $S=(P,B,I)$, with an incidence relation satisfying the following axioms

(i) each point is incident with r lines ($r \geqslant 2$) and two distinct points are incident with at most one line;

(ii) each line is incident with k points ($k \geqslant 2$) and two distinct lines are incident with at most one point;

(iii) if x is a point and L is a line not incident with x, then there are a unique point x' and a unique line L' such that xIL'Ix'IL.

If $|P| = v$ and $|B| = b$, then $v=k(kr-k-r+2)$ and $b=r(kr-k-r+2)$. In [7] D.G.Higman proves that the positive integer $k+r-2$ divides $kr(k-1)(r-1)$. Moreover, under the assumption that $k>2$ and $r>2$, he shows that $r \leqslant (k-1)^2+1$ and $k \leqslant (r-1)^2+1$.

1.2. Let $P=\{x_{ij} \| i,j=1,2,\ldots,k\}$ and $B=\{L_1,L_2,\ldots,L_k,M_1,M_2,\ldots,M_k\}$, where $k \geqslant 2$. Incidence is defined as follows : $x_{ij}IL_1 \Longleftrightarrow i=1, x_{ij}IM_1 \Longleftrightarrow j=1$. Then $S=(P,B,I)$ is a 4-gonal configuration with parameters
$$k=k, r=2, v=k^2, b=2k.$$

1.3. (J.Tits [4]) We consider a non-singular hyperquadric Q of index 2 of the projective space $PG(d,q)$, with $d=3,4$ or 5. Then the points of Q together with the lines of Q (which are the subspaces of maximal dimension on Q) form a 4-gonal configuration with parameters
$$k=q+1, r=2, v=(q+1)^2, b=2(q+1), \text{when } d=3;$$
$$k=r=q+1, v=b=(q+1)(q^2+1), \text{when } d=4;$$
$$k=q+1, r=q^2+1, v=(q+1)(q^3+1), b=(q^2+1)(q^3+1),$$
$$\text{when } d=5.$$

1.4. (J.Tits [4]) Let H be a non-singular Hermitian primal [11] of the projective space PG(d,q),q=p^{2h}.If d=3 or 4,then the points of H together with the lines of H form a 4-gonal configuration with parameters

for d=3 : k=q+1,r=1+$\sqrt{q}$,v=(q+1)(1+q$\sqrt{q}$),b=(1+$\sqrt{q}$)(1+q$\sqrt{q}$);

for d=4 : k=q+1,r=1+q$\sqrt{q}$,v=(q+1)(1+$q^2\sqrt{q}$),b=(1+q$\sqrt{q}$)(1+$q^2\sqrt{q}$).

1.5. (J.Tits [4]) Let d=2 (resp. d=3) and consider a (q+1)-arc (*) K (resp. a (q^2+1)-cap (*) K) of PG(d,q) (not at the same time d=3 and q=2).Let PG(d,q) be embedded as a hyperplane H in PG(d+1,q)=P.Define points as (i) the points of P-H,(ii) the hyperplanes X of P for which $|X \cap K|$=1,and (iii) one new symbol u.Lines are (a) the lines of P which are not contained in H and meet K (necessarily in a unique point),and (b) the points of K.Incidence is defined as follows : Points of type (i) are incident only with lines of type (a);here the incidence is that of P.A point X of type (ii) is incident with all lines $\subset$X of type (a) and with precisely one line of type (b),namely the one represented by the unique point of K in X.Finally,the unique point u of type (iii) is incident with no line of type (a) and all lines of type (b).The incidence structure so defined is a 4-gonal configuration with parameters

k=r=q+1,v=b=(q+1)(q^2+1),when d=2;

k=q+1,r=q^2+1,v=(q+1)(q^3+1),b=(q^2+1)(q^3+1),when d=3.

1.6. REMARK. The points of PG(3,q),together with the totally isotropic lines with respect to a symplectic polarity π , form a 4-gonal configuration W(q) with parameters k=r=q+1, v=b=(q+1)(q^2+1).This configuration is the dual of the 4-gonal configuration arising from a non-singular hyperquadric of index 2 in PG(4,q) [2] .

(*) A t-arc of PG(2,q) (resp. a t-cap of PG(3,q)) is a set of t points of PG(2,q) (resp. PG(3,q)),no three of which are collinear.

1.7. (S.E.Payne [9]) Consider a 4-gonal configuration S
with parameters $k=r=s+1, v=b=(s+1)(s^2+1)$. If x and y are distinct points of S, let $z_0, z_1, \ldots, z_s$ be the points collinear
with both x and y. Then the pair (x,y) is called regular
provided that any point collinear with two of the z_i's is
collinear with all of them. The point x is regular if the
pair (x,y) is regular for all points $y, y \neq x$ (for example,
every point of the 4-gonal configuration W(q) is regular).
Dual definitions for regular pairs of lines and regular
lines.

Let x be any regular point of S, and let L_1 and L_2 be two
lines through x. A 4-gonal configuration S* is defined as
follows. Points of S* are precisely the points of S which
are not collinear in S with x. The lines of S* are of two
types. Those of type (i) are the lines of S not incident
with x, and those of type (ii) are indexed by pairs $[x_1, x_2]$
of points such that $x_i IL_i, x_i \neq x, i=1,2$. For lines of type (i)
the incidence is that of S restricted to points of S*. The
line $[x_1, x_2]$ is to be incident (in S*) with precisely those
points of S different from x which are collinear in S with
both x_1 and x_2. The incidence structure so defined is a 4-go-
nal configuration with parameters
$$k=s, r=s+2, v=s^3, b=s^3+2s^2.$$
There follows that there exists a 4-gonal configuration
with parameters
$$k=q, r=q+2, v=q^3, b=q^3+2q^2, \text{where q is any prime}$$
power.

REMARK. For particular cases see [1] and [6].

1.8. (J.A.Thas [16]) In PG(3n-1,q), $q=2^h$ and $n \geqslant 1$, we consider q^n+2 (n-1)-dimensional subspaces $PG^{(1)}(n-1,q), \ldots,$
$PG^{(q^n+2)}(n-1,q)$, every three of them being joined by
PG(3n-1,q) (with such a set of subspaces there corresponds
a (q^n+2)-arc of the projective plane over the total matrix
algebra of the n x n - matrices with elements in GF(q) [14]).

Let $PG(3n-1,q)$ be embedded as a hyperplane H in $PG(3n,q)=P$. Define points of the 4-gonal configuration as the points of P-H.Lines are the n-dimensional subspaces of P through $PG^{(i)}(n-1,q)$,$i=1,2,\ldots,q^n+2$,which are not contained in H. Incidence is that of P.

The incidence structure so defined is a 4-gonal configuration with parameters
$$k=q^n,r=q^n+2,v=q^{3n},b=q^{3n}+2q^{2n} \quad (q=2^h,n\geqslant1).$$

THE PARTICULAR CASE n=1. For n=1 we obtain a 4-gonal configuration with parameters $k=q,r=q+2,v=q^3,b=q^3+2q^2$ $(q=2^h)$, arising from a (q+2)-arc in $PG(2,q)$ (so we find again the model constructed by M.Hall,Jr.[6]).

1.9. (J.A.Thas [16]) In $PG(3n-1,q)$,$q=p^h$ and $n\geqslant1$,we consider q^n+1 (n-1)-dimensional subspaces $PG^{(1)}(n-1,q),\ldots,$ $PG^{(q^n+1)}(n-1,q)$,every three of them being joined by $PG(3n-1,q)$ (with such a set of subspaces there corresponds a (q^n+1)-arc K of the projective plane over the total matrix algebra of the n x n - matrices with elements in GF(q) [14]).In [14] we have proved that through $PG^{(i)}(n-1,q)$, $i=1,2,\ldots,q^n+1$,there passes one and only one subspace $PG^{(i)}(2n-1,q)$ of $PG(3n-1,q)$ which has no point in common with the set $PG^{(1)}(n-1,q)\cup\ldots\cup PG^{(i-1)}(n-1,q)\cup PG^{(i+1)}(n-1,q)\cup\ldots\cup PG^{(q^n+1)}(n-1,q)$ (with the q^n+1 spaces $PG^{(i)}(2n-1,q)$ there correspond the q^n+1 tangent lines of the (q^n+1)-arc K [14]).

Let $PG(3n-1,q)$ be embedded as a hyperplane H in $PG(3n,q)=P$.Define points of the 4-gonal configuration as (i) the points of P-H (ii) the 2n-dimensional subspaces X of P for which $X\cap H=PG^{(i)}(2n-1,q)$,$i\in\{1,2,\ldots,q^n+1\}$ (iii) one new symbol u.Lines of the configuration are (a) the n-dimensional subspaces of P which are not contained in H and pass through one of the spaces $PG^{(1)}(n-1,q),PG^{(2)}(n-1,q),\ldots,$ $PG^{(q^n+1)}(n-1,q)$,and (b) the spaces $PG^{(1)}(n-1,q),\ldots,$

$PG^{(q^n+1)}(n-1,q)$.Incidence is defined as follows : Points of
type (i) are incident only with lines of type (a) ; here the
incidence is that of P.A point X of type (ii) is incident
with all lines $\subset X$ of type (a) and with precisely one line
of type (b),namely the one represented by the unique space
$PG^{(i)}(n-1,q)$ in X.Finally,the unique point u of type (iii)
is incident with no line of type (a) and all lines of type
(b).

The incidence structure so defined is a 4-gonal configuration with parameters
$$k=r=q^n+1,v=b=(q^n+1)(q^{2n}+1) \quad (q=p^h \text{ and } n \geqslant 1).$$

THE PARTICULAR CASE n=1. For n=1 we obtain a 4-gonal configuration with parameters $k=r=q+1,v=b=(q+1)(q^2+1)$ $(q=p^h)$,
arising from a (q+1)-arc in PG(2,q) (so we find again one
of the models constructed by J.Tits).

1.10. (J.A.Thas [16]) In PG(4n-1,q),$q=p^h$ and $n \geqslant 1$ (and not
at the same time q=2,n=1),we consider $q^{2n}+1$ (n-1)-dimensio-
nal subspaces $PG^{(1)}(n-1,q),...,PG^{(q^{2n}+1)}(n-1,q)$,satisfying
the following

(i) every three of these spaces are joined by a (3n-1)-
dimensional subspace of PG(4n-1,q);

(ii) every four of these spaces are contained in a
PG(3n-1,q) or are joined by PG(4n-1,q);
(with such a set of subspaces there corresponds an ovaloid
K of the threedimensional projective space over the total
matrix algebra of the n x n - matrices with elements in
GF(q) [14]).In [14] we have proved that through $PG^{(i)}(n-1,q)$,
i=1,2,...,$q^{2n}+1$,there passes one and only one subspace
$PG^{(i)}(3n-1,q)$ of PG(4n-1,q) which has no point in common
with the set $PG^{(1)}(n-1,q) \cup...\cup PG^{(i-1)}(n-1,q)\cup PG^{(i+1)}(n-1,$
$q)\cup...\cup PG^{(q^{2n}+1)}(n-1,q)$ (with the $q^{2n}+1$ spaces $PG^{(i)}(3n-1,$
q) there correspond the $q^{2n}+1$ tangent planes of the
$(q^{2n}+1)$-cap K [14]).

Let PG(4n-1,q) be embedded as a hyperplane H in PG(4n,q)=

=P.Define points of the 4-gonal configuration as (i) the points of P-H (ii) the 3n-dimensional subspaces X of P for which $X \cap H = PG^{(i)}(3n-1,q), i \in \{1,2,\ldots,q^{2n}+1\}$ (iii) one new symbol u.Lines of the configuration are (a) the n-dimensional subspaces of P which are not contained in H and pass through one of the spaces $PG^{(1)}(n-1,q), PG^{(2)}(n-1,q),\ldots,$ $PG^{(q^{2n}+1)}(n-1,q)$,and (b) the spaces $PG^{(1)}(n-1,q),\ldots,$ $PG^{(q^{2n}+1)}(n-1,q)$.Incidence is defined as follows : Points of type (i) are incident only with lines of type (a) ; here the incidence is that of P.A point X of type (ii) is incident with all lines $\subset X$ of type (a) and with precisely one line of type (b),namely the one represented by the unique space $PG^{(i)}(n-1,q)$ in X.Finally,the unique point u of type (iii) is incident with no line of type (a) and all lines of type (b).

The configuration S so defined is a 4-gonal configuration with parameters
$$k=q^n+1, r=q^{2n}+1, v=(q^n+1)(q^{3n}+1), b=(q^{2n}+1)(q^{3n}+1)$$
$(q=p^h, n \geqslant 1,$ and not at the same time $q=2, n=1)$.

THE PARTICULAR-CASE n=1. For n=1 we obtain a 4-gonal configuration with parameters $k=q+1, r=q^2+1, v=(q+1)(q^3+1),$ $b=(q^2+1)(q^3+1)$ $(q=p^h$ and $q>2)$,arising from a (q^2+1)-cap in PG(3,q) (so we find again one of the models constructed by J.Tits).

1.11. OPEN QUESTION. Do the general constructions 1.8.,1.9., 1.10. yield configurations which are not isomorphic to configurations provided by the particular cases n=1 ?

2. 4-gonal configurations of order n

2.1. Definitions

A 4-gonal configuration of order n is a 4-gonal configuration with parameters $r=k=n+1, v=b=(n+1)(n^2+1)$.

Let $S=(P,B,I)$ be a 4-gonal configuration of order n and

let x and y be distinct points of S.The trace of x and y is
defined to be the set tr$\{x,y\}$={all z $\in$ P ‖ z is collinear
with both x and y}.If $(x,y),x,y\in$P,is a regular pair of S
and if z_1 and z_2 are distinct points of tr$\{x,y\}$,then
tr$\{z_1,z_2\}$ is said to be the span of x and y (notation :
sp$\{x,y\}$).

2.2. Theorem (R.R.Singleton,C.T.Benson,S.E.Payne)

Let x be a regular point of a 4-gonal configuration S
of order n.Let π_x be the incidence structure whose points
are the points of S collinear with x and whose lines are
the spans of the (necessarily regular) pairs of distinct
points of π_x.Then π_x with the natural incidence relation
is a projective plane of order n.

2.3. Theorem (J.A.Thas)

Consider a 4-gonal configuration S=(P,B,I) of order n
(n $\geqslant$ 2),and suppose that K is a pointset of S satisfying the
following

(i) $|K|=n^2+1$;

(ii) every line of S is incident with one and only one
point of K (i.e. no two points of K are collinear) ;

(iii) for any three distinct points x,y,z of K there is
some point w collinear with all of x,y,z.

Then $n=2^h$ and S=(P,B,I) is the 4-gonal configuration W(n)
arising from a symplectic polarity π of PG(3,n) (see 1.6.).
Moreover K is a (n^2+1)-cap of PG(3,n),for which the tangent
lines coincide with the totally isotropic lines of the polarity π.

Proof. First of all we define the following incidence
structure $I^*(K)$=(K,B*,I*):

the points of $I^*(K)$ are the elements of K ;

the blocks of $I^*(K)$ (i.e. the elements of B*) are called
circles and are defined as follows : the n+1 lines incident
with an arbitrary point p $\in$ P-K are incident with n+1 points

J.A.THAS

of K.The set defined by these n+1 points is a circle C of
I*(K) ;

if $x \in K$ and $C \in B^*$, then xI^*C if and only if $x \in C$.

Now we show that I*(K) is an inversive plane of order n.

a) The number of circles is at most $|P-K|=n^3+n$. From (iii)
there follows that through every three points of K there
passes at least one circle. So the number of circles is at
least $\frac{(n^2+1)n^2(n^2-1)}{(n+1)n(n-1)} = n^3+n$. Consequently $|B^*|=n^3+n$. Hence it
appears that through every three points of K there passes
óne and only one circle.

From the preceding there also follows that every circle
C is defined by one and just one point $p \in P-K$. We say that p
is the nucleus of C and that the lines incident with p are
the tangent lines of C.

b) Consider a circle C,a point $x \in C$ and a point $y \in K-C$.
We shall prove that there is a unique circle C',with $y \in C'$
and $C \cap C'=\{x\}$.

First of all we remark that the number of circles,dif-
ferent from C,through x equals $|\{$all $p \in P-K \| \exists L \in B$ with
$xILIp\}|-1 = n(n+1)-1 = n^2+n-1$. Next we determine the number
of circles which pass through the points x and $z \in C$, where
$z \neq x$. This number equals $|\{$all $p \in P-K \| \exists L,L' \in B$ with
$xILIpIL'Iz\}| = n+1$ (to obtain this result, take an arbitrary
line L incident with x; from axiom (iii) in the definition
of a 4-gonal configuration it follows that there is a unique point $p \in P-K$ and a unique line L' such that $zIL'IpIL$;
so the number we are looking for equals the number of lines
L incident with x). Now it is easy to deduce that there are
exactly n-1 circles C',with $C \cap C'=\{x\}$.

Next we consider the nucleus p of C, and the tangent line
L of C which is incident with x.If qIL and $q \notin \{x,p\}$, then we
prove that the circle C' with nucleus q is tangent to C.
Suppose that $C \cap C'=\{x,z\}$ and that L' (resp. L") is the tangent line of C (resp. C') at the point z.Then $zIL'IpIL$ and

zIL"IqIL,in contradiction with axiom (iii) in 1.1.So there
results that $C \cap C' = \{x\}$.In this way we obtain the n-1 circles C',with $C \cap C' = \{x\}$.Finally,we consider the unique point
q and the unique line **M**,defined by yIMIqIL $(q \notin \{x,p\})$.Then
the circle C' with nucleus q evidently is the unique circle
with $y \in C'$ and $C \cap C' = \{x\}$.

c) $|K| \geqslant 5, |C| > 0$ for all $C \in B^*$,and there exist a $x \in K$
and a $C \in B^*$ with $x \notin C$.

From (a),(b) and (c) there follows immediately that I*(K)
is an inversive plane of order n.Next we prove that n is
even.

For that purpose we consider a circle C,a point $x \in C$ and
a point $y \in K-C$.Through y there passes a unique circle C',
with $C \cap C' = \{x\}$.Now we take an arbitrary point $z \in \hat{C}$,with
$z \neq x$,and we consider the unique circle C",with $z \in C''$ and
$C' \cap C'' = \{y\}$.We shall prove that $|C \cap C''| = 2$.Suppose a moment
that $C \cap C'' = \{z\}$.The nucleus of C (resp. C',C") is denoted by
p (resp. p',p"),and the tangent line of C' or C" (resp. C"
or C,C or C') at the point y (resp. z,x) is denoted by L
(resp. L',L").Then pⴑL,pIL"Ip'IL and pIL'Ip"IL,in contradic-
tion with axiom (iii) in 1.1.Consequently $|C \cap C''| = 2$ (we al-
so remark that $x \notin C \cap C''$).It follows immediately that $|C| - 1 =$
$= n$ is even,and so I*(K) is an egglike inversive plane of
even order n [3,4] .There results that n is of the form 2^h.

Since I*(K) is an egglike inversive plane of order $n=2^h$
there exist a (n^2+1)-cap K' in PG(3,n) and a bijection σ
of K' onto K,such that for every plane H of PG(3,**n**) with
$|H \cap K'| > 1, (K' \cap H)^\sigma$ is a circle of I*(K).If $W(n)=(P',B',\in)$
is the 4-gonal configuration arising from the symplectic
polarity π defined by K',then we define as follows the bijection φ of $P' \cup B'$ onto $P \cup B$:

(i) $x^\varphi = x^\sigma$,when $x \in K'$;

(ii) let $x \in P'-K'$ and call H the polar plane of x with
respect to the symplectic polarity π;then x^φ is the nucleus of the circle $(K' \cap H)^\varphi$ of I*(K) ;

(iii) let $L \in B'$ and suppose that L contains the point x
of K';if H is a plane with $L \subset H$ and $|H \cap K'| > 1$, then L^{ψ} is
the tangent line of $(H \cap K')^{\psi}$ at the point x^{ψ}.

From the preceding there follows immediately that ψ is
a bijection of $P' \cup B'$ onto $P \cup B$ and that $x \in L \Longleftrightarrow x^{\psi} \text{ IL}^{\psi}$,
for all $x \in P'$ and $L \in B'$.Consequently ψ is an isomorphism
of $W(2^h)$ onto $S=(P,B,I)$,with $K'^{\psi} = K$.And so our theorem is
completely proved.

Corollaries

a) If q is an odd prime power then the maximal number of
points of $W(q)$,no two of which are collinear,is less than
$1+q^2$.

b) Consider a 4-gonal configuration $S=(P,B,I)$ of order
n $(n \geqslant 2)$,and suppose that K is a pointset of S satisfying
the following

(i) $|K|=n^2+1$;

(ii) no two points of K are collinear ;

(iii) every point of K is regular.

Then $n=2^h$ and $S=(P,B,I)$ is the 4-gonal configuration
$W(n)$ arising from a symplectic polarity π of $PG(3,n)$.More-
over K is a (n^2+1)-cap of $PG(3,n)$,for which the tangent lines coincide with the totally isotropic lines of the polarity π.

Proof. It is sufficient to prove that for any three distinct points x,y,z of K there is some point w of S collinear with all of x,y,z.So we consider three distinct points
x,y,z of K.Then $\text{tr}\{x,y\}$ and $\text{tr}\{x,z\}$ are two lines of the
projective plane π_x defined by x (see 2.2.).The common
point w of these two lines of π_x evidently is collinear
with all of x,y,z.

2.4. Theorem (J.A.Thas)

Suppose that the 4-gonal configuration $S=(P,B,I)$ of order n has a regular point x and a regular pair of lines

J.A.THAS

(Y,Z),where Y and Z are not concurrent.If no line of tr{Y,Z}
is incident with x,then n is even.

Proof. Consider the set Q={all $p \in P \| \exists L \in tr\{Y,Z\}$ with pIL}
={all $p \in P \| \exists L \in sp\{Y,Z\}$ with pIL} (we remark that $x \notin Q$).The
set Q contains exactly n+1 points of the projective plane
π_x defined by the regular point x (each of these points is
incident with an element of tr{Y,Z} (resp. sp{Y,Z});conver-
sely,each line of tr{Y,Z}(resp. sp{Y,Z}) is incident with
one of these n+1 points).As $x \notin Q$ the set K of these n+1
points does not contain x.Taking account of the regularity
of the pair (Y,Z) it is easy to show that each line of π_x
through x contains exactly one point of the set K (if a line
of π_x through x would contain at least two points of K,then
there follows immediately that $x \in Q$).

Next we shall prove that K is a (n+1)-arc of the projective plane π_x.Suppose that the points u,v,w of K are collinear in the plane π_x.Let U be the line of tr{Y,Z} which
is incident with u,and let V (resp. W) be the line of sp{Y,Z}
which is incident with v (resp. w).If a and b are the points
defined by UIaIV and UIbIW,then evidently a≠b.As u,v,w are
collinear in π_x,we have $b \in tr\{u,v\}$.There follows immediately that a=b,a contradiction.So we conclude that K is a
(n+1)-arc of the projective plane π_x.

From the preceding there follows immediately that the
tangents of the (n+1)-arc K all meet in x.Hence it appears
that n is even.

Corollaries

a) If the 4-gonal configuration S=(P,B,I) of order n has
a regular point x and a regular line Y,with x∤Y,then n is
even.

Proof. Let x' and Y' be defined by xIY'Ix'IY.If Z is a
line which is not concurrent with Y or Y',then (Y,Z) is a
regular pair of lines.Moreover Y and Z are not concurrent
and no line of tr{Y,Z} is incident with x.From the theorem

there follows immediately that n is even.

b) If the 4-gonal configuration S=(P,B,I) of order n,n odd,contains two regular points,then S is not self-dual.

Proof. Let x and y be two regular points of S and suppose that g is an anti-automorphism of S.Then at least one of the regular lines x^g, y^g is not incident with at least one of the regular points x and y.From (a) there follows immediately that n is even,a contradiction.

c) The 4-gonal configuration W(q),q=p^h and p odd, is never self-dual (see also [2] and [15]).

Proof. Every point of W(q) is regular,and so from (b) there follows immediately that W(q),q odd,is never self-dual.

REFERENCES

1. R.W.AHRENS and G.SZEKERES,On a combinatorial generalization of 27 lines associated with a cubic surface,J. Austral. Math. Soc.,X (1969),485-492.

2. C.T.BENSON,On the structure of generalized quadrangles, J. Algebra,15 (1970),443-454.

3. P.DEMBOWSKI and D.R.HUGHES,On finite inversive planes, J. London Math. Soc.,40 (1965),171-182.

4. P.DEMBOWSKI,"Finite geometries," Springer-Verlag,1968, 375 pp.

5. W.FEIT and D.G.HIGMAN,The nonexistence of certain generalized polygons,J. Algebra,1 (1964),114-131.

6. M.HALL,Jr.,Affine generalized quadrilaterals,Studies in Pure Mathematics,ed. by L.Mirsky,Academic Press,1971, 113-116.

7. D.G.HIGMAN,Partial geometries,generalized quadrangles and strongly regular graphs,Atti del convegno di geometria combinatoria e sua applicazioni,Perugia (1971).

8. S.E.PAYNE,Affine representations of generalized quadran-

gles,J. Algebra,16 (1970),473-485.

9. S.E.PAYNE,Nonisomorphic generalized quadrangles,J. Algebra,18 (1971),201-212.

10.S.E.PAYNE,The equivalence of certain generalized quadrangles,J. Comb. Theory,10 (1971),284-289.

11.B.SEGRE,Introduction to Galois geometries,Atti della Accad. Naz. Lincei,Mem. Cl. di Sc. Fis. Mat, e Nat.,8, Sez. I^a,Fasc. 5 (1967),137-236.

12.R.R.SINGLETON,Minimal regular graphs of maximal even girth,J. Comb. Theory,1 (1966),306-332.

13.J.A.THAS,Een studie betreffende de projectieve rechte over de totale matrix algebra $M_3(K)$ der 3 x 3 - matrices met elementen in een algebraïsch afgesloten veld K,Verh. Kon. Vl. Acad. Wet.,Lett. Sch. K. van Belgiü,Kl. der Wet, 31,no. 112 (1969),151 pp.

14.J.A.THAS,The m-dimensional projective space $S_m(M_n(GF(q)))$ over the total matrix algebra $M_n(GF(q))$ of the n x n - matrices with elements in the Galois field GF(q),Rend. di Mat.,(6) 4 (1971),459-532.

15.J.A.THAS,Ovoidal translation planes,Arch. der Math., (23) 1 (1972),110-112.

16.J.A.THAS,On 4-gonal configurations,Geometriae Dedicata (to appear).

17.J.TITS,Sur la trialité et certains groupes qui s'en déduisent,Publ. Math. I.H.E.S. Paris,2 (1959),14-60.

18.J.TITS,Géométries polyédriques et groupes simples,2e Reunion Math. d'expression latine (Firenze-Bologna 1961), Cremonese,Roma,66-88.

19.J.TITS,Géométries polyédriques finies,Rend. di Mat.,23 (1964),156-165.

CENTRO INTERNAZIONALE MATEMATICO ESTIVO
(C. I. M. E.)

H. P. YOUNG

AFFINE TRIPLE SYSTEMS

Corso tenuto a Bressanone dal 18 al 27 Giugno 1972

AFFINE TRIPLE SYSTEMS

H. P. Young

1. Matroids

In this paper we present a new class of finite geometries
(matroids) that have the same structural regularity found in the classical
affine and projective geometries, and whose coördinatization by a class
of Moufang loops provides illuminating links between geometric and
algebraic concepts. A more detailed account of this work may be found
in [9].

A matroid $M = (E, \mathcal{Q})$ is a finite set E together with a non-
empty set $\mathcal{Q}$ of subsets of E, called independent sets, and satisfying
the following axioms of independence. (I1) Every subset of an independ-
ent set is independent; and (I2) For every $A \subseteq E$, every two maximal
independent subsets of A have the same cardinality, called the rank of
A, $r(A)$.

For any $A \subseteq E$, the submatroid of M on A, denoted by $M \times A$, is
the matroid on set A with independent sets $\mathcal{Q}_A = \{J \in \mathcal{Q} : J \subseteq A\}$.
A k-flat is a maximal set having rank k. The hyperplanes are the
$(r(E)-1)$-flats. $\mathcal{H}$ is the hyperplane family of a matroid on E if and
only if: (H1) The members of $\mathcal{H}$ are proper subsets of E and no
member of $\mathcal{H}$ is a proper subset of another; and (H2) For any distinct
H_1, H_2 in $\mathcal{H}$ and $x \in E$, $(H_1 \cap H_2) \cup \{x\}$ is contained in some member
of $\mathcal{H}$.

A matroid design is a matroid in which all the hyperplanes have
the same cardinality. A perfect matroid design (PMD) is a matroid such
that for every k, all k-flats have the same cardinality, $\alpha(k)$. [See [7]

H. P. Young

and [8])

A Steiner triple system $S = (E, \mathcal{L})$ with $\lambda = 1$ is an example of
a rank 3 PMD, whose hyperplane family $\mathcal{L}$ is just the set of blocks of S.
We will call the 1-flats of S __points__ and the 2-flats __lines__. The
maximal independent subsets of E, called __bases__ of S, are just the non-
collinear triples. A subset $A \subseteq E$ is __2-closed__ if for every $x, y \in A$
the unique line containing x and y is contained in A. The __2-closure__
of A, $cl_2(A)$ is the intersection of all 2-closed sets containing A.
The 2-closure of a basis will be called a __plane__ of S. If every plane
of S is an (order 3) affine plane, then we call S an __affine triple
system__. If $S = (E, \mathcal{H})$ is an affine triple system and $\mathcal{P}$ is the family
of planes of S, it is readily verified by (H1) and (H2) that $(E, \mathcal{P})$ is
a rank 4 PMD in which $\alpha(1) = 1$, $\alpha(2) = 3$, $\alpha(3) = 9$, and $\alpha(4) = |E|$.
In [3], M. Hall constructs an affine triple system whose points and lines
are not isomorphic to the points and lines of an affine geometry of order
3; i.e., whose 2-closed subsets are not the subspaces of some $AG(n,3)$.
The rank 4 PMD associated with any such affine triple system will be
called a __Hall__ __matroid__. In this paper we will show that Hall matroids can
be coördinatized by a class of Moufang loops, and by this method we are
able to construct an infinite class of them and analyze their geometric
properties.

2. Coördinatization by Loops

Bruck [1] seems to have been the first to note several ways in
which triple systems may be coordinatized by loops. (See also [4]) Here
we develop some of these ideas further.

H. P. Young

For every Steiner triple system $S = (E, \mathscr{L})$ we may make E into a quasigroup by defining a binary operation "$\circ$" as follows.

$$x \circ x = x \qquad \text{(idempotence)}$$

For $x \neq y$, $x \circ y = z$ if and only if $\{x,y,z\}$ is a triple. Then "$\circ$" is commutative and furthermore satisfies

$$x \circ (x \circ y) = y \quad \text{for all} \quad x,y \in E .$$

However, there are more useful ways of associating quasigroups, and in particular, loops, with a given triple system $S = (E, \mathscr{L})$. Let "$\circ$" always denote the binary operator defined above on the points of the triple system. For a given $e \in E$, define the <u>loop of</u> S <u>based at</u> e, $G_e(S)$, as follows.

(1) For all $x,y \in E$, $\quad xy = (e \circ x) \circ (e \circ y)$.

It then follows from the properties of "$\circ$" that:

(i) $xe = ex = x$ for all $x \in E$ (e is the identity of $G_e(S)$;

(ii) $xx = e \circ x$ is the unique inverse of x, since

$(e \circ x) \circ x = x \circ (e \circ x) = e$; (iii) $xy = yx$ for all x and y.

Thus $G_e(S)$ is a commutative loop of exponent 3. Note that there is nothing sacred about this way of associating a loop with S. For example, we could just as well have defined $xy = e \circ (x \circ y)$.

In view of (ii) above we shall sometimes write x^{-1} for x^2. Thus xy is the third point of the unique triple containing x^{-1} and y^{-1}, or

(2) $$xy = x^{-1} \circ y^{-1} .$$

Now can we, given a commutative loop G of exponent 3 with identity e, reverse the process and construct a triple system whose

H. P. Young

loop based at e is G? In general, we cannot. Indeed, the triple containing distinct elements x^{-1} and y^{-1} would have to be $\{x^{-1}, y^{-1}, xy\}$, and so we must also have

$$x^{-1} = (y^{-1})^{-1}(xy)^{-1}$$
and
$$y^{-1} = (xy)^{-1}(x^{-1})^{-1}$$

for all x and y .

Since $(x^{-1})^{-1} = x$ in a commutative loop, the above relations are equivalent to

$$y(xy)^{-1} = x^{-1} \qquad \text{for all } x,y \in G ,$$

which is known as the __weak__ __inverse__ __property__ for a commutative loop [6].

(3)   Thus, if G is a commutative loop of exponent 3 with the weak inverse property and we let $\mathscr{L} = \{\{x^{-1}, y^{-1}, xy\} : x,y \in G, \ x \neq y\}$, then $(G,\mathscr{L})$ is a triple system whose loop based at e is G.

A basic result relating the structure of a triple system to that of its coördinatizing loop is the following.

(4)   __Theorem__. Let $S = (E, \mathscr{L})$ be a Steiner triple system, $|E| > 3$. For any $e \in E$, $G_e(S)$ is a group (in particular, an elementary abelian 3-group) if and only if the 2-closed subsets of E are the subspaces of an $AG(n,3)$ for some $n \geq 2$.

3. Hall Matroids

We may characterize affine triple systems $(E, \mathscr{L})$ by the following incidence axiom.

(A)   If $L = \{x,y,z\} \in \mathscr{L}$ and $w \notin L$, then $\{w \circ x, w \circ y, w \circ z\} \in \mathscr{L}$.

H. P. Young

The proof is obtained at once by verifying that the 2-closure of any basis of $(E, \mathcal{L})$ has exactly nine points. A proof is also given by M. Hall [3] for the following equivalent criterion.

(5)   Theorem. (Hall) S is an affine triple system if and only if for every point w there is an involutory automorphism of S fixing exactly w.

Now let us consider the implications of Axiom (A) for the coördinatizing loop.

A loop G is said to be a <u>Moufang</u> loop if for all x,y,z ∈ G

$$(xy)(zx) = (x(yz))x$$

Then we may prove:

(6)   Theorem. Let $S = (E, \mathcal{L})$ be a Steiner triple system.

   (i) If S is affine, then the loops $G_e(S)$, e ∈ E, are
        Moufang, and they are all isomorphic.

   (ii) If for some e ∈ E, $G_e(S)$ is Moufang, then S is affine.

Since every Moufang loop has the weak inverse property, (3), (4), and (6) establish a one-to-one correspondence between Hall matroids and commutative Moufang loops of exponent 3 (abbreviated hereafter by "c.M. 3-loops"). Thus we have available Bruck's powerful theory on commutative Moufang loops [2] for analyzing the structure of such systems and for constructing new ones. A basic result is Moufang's theorem [5].

(7)   Theorem. (Moufang) Let G be a Moufang loop.

   (i) G is di-associative, i.e., every two elements generate

H. P. Young

a subgroup.

.   (ii) If (ab)c = a(bc) for some a,b,c ∈ G then a,b,c

generate a subgroup of G.

Let G be a commutative Moufang loop. The **associator** of x,y,

and z is the unique element (x,y,z) of G such that

(8)   (xy)z = [x(yz)](x,y,z)

The associator is skew-symmetric in its arguments,

(9)   (x,y,z) = (z,x,y) = (y,x,z)$^{-1}$

and satisfies the relations

(10)   (x^{-1},y,z) = (x,y,z)$^{-1}$

and

(11)   (x,xy,z) = (x,y,z) .

We also have the following useful expansion formula

(12)   (wx,y,z) = [(w,y,z)(x,y,z)][((w,y,z),w,x)((x,y,z),x,w)]

The **center** of G is the subloop

Z = {a ∈ G: ∀ x,y ∈ G, ax = xa and (a,x,y) = (x,a,y) = (x,y,a) = e}

H is **normal** in G (H ◄ G) if for all x,y ∈ G, xH = Hx,

x(yH) = (xy)H, and (Hy)x = H(yx). If H ◄ G then the quotient loop is

well-defined.

(13)   **Theorem**. ([2], Ch. VII, Lemma 5.5) If G is a commutative

Moufang loop then (x,y,(x,y,z)) = e, and, more generally, (x,y,z) is

in the center of the subloop generated by x,y, and z.

H. P. Young

The Hall matroid found by Hall [3] had 81 points and was constructed by an argument using automorphism groups. Using our point of view we can prove that Hall's example is the unique one of this order.

(14)   Theorem. There is a unique c.M. 3-loop on three generators that is not a group, and it has order 81. Further, this loop and the elementary abelian 3-group on four generators are the only c.M. 3-loops of order 81.

Proof. Let G be a c.M. 3-loop, not a group, generated by a, b, c and with identity e. Let $h = (a,b,c) \neq e$, and $H = \langle h \rangle$ the subgroup generated by h. By (13), $H \lhd G$, so G/H is defined and generated by aH, bH, cH, which satisfy $(aH \cdot bH) \cdot cH = aH \cdot (bH \cdot cH)$. By Moufang's theorem (7) G/H is therefore a group. In particular, G/H is an abelian 3-group on at most three generators. If G/H has less than three generators, then without loss of generality we could write $c = h(ab)$. But then $(ab)c = (ab)(h(ab)) = h(a^2b^2) = a(bc)$, by di-associativity. Hence, again by (7), G is a group. This contradiction shows that $|G/H| = 27$, and so $|G| = 81$. Where Z is the entire center of G $(H \subseteq Z)$, a similar argument shows that $|G/Z| = 27$, whence $Z = H$. Further, it follows that any three non-associating elements of G generate G, so

(15) $(x,y,z) \in Z$ for all x,y,z in G.

Hence the expansion formula (12) may be written more simply as follows:

(16) $(wx,y,z) = (w,y,z)(x,y,z)$

H. P. Young

Since we know, by Hall's construction, that there exists such a loop G, we need only show that its multiplication is uniquely determined. Letting $A = \{e,a,a^{-1}\}$, $B = \{e,b,b^{-1}\}$, $C = \{e,c,c^{-1}\}$, the preceding shows that every element of G is represented by a member of the set $H(A(BC)) = K$, and this representation is unique because $|K| = 81$. For any two elements $x = h(x_1(x_2x_3))$ and $x' = h'(x_1'(x_2'x_3'))$ of K we have

$$xx' = hh'[x_1(x_2x_3)][x_1'(x_2'x_3')]$$

$$= hh'h_1x_1[(x_2x_3)(x_1'(x_2'x_3'))] ,$$

where $h_1 = (x_1,x_2x_3,x_1'(x_2'x_3'))$ is in the center of G. Continuing in this manner we obtain

$$xx' = hh'h_1h_2 \quad x_1[x_1'((x_2'x_3')(x_2x_3))]$$

$$= hh'h_1h_2h_3(x_1x_1')[(x_2'x_3')(x_2x_3)]$$

where $\quad h_2 = (x_1',x_2'x_3',x_2x_3)$

and

$$h_3 = (x_1,x_1',(x_2'x_3')(x_2x_3))^{-1} .$$

But $(x_2'x_3')(x_2x_3) = (x_2'x_2)(x_3'x_3)$, by di-asociativity, hence

$$xx' = hh'h_1h_2h_3[(x_1x_1')[(x_2x_2')(x_3x_3')]] \in K .$$

By (16) and permutations thereof, we find that h_1, h_2, h_3 are uniquely determined by x and x'. Hence xx' is a uniquely determined member of K.

Conversely, let G be any c.M. 3-loop of order 81, and let $k(G)$ be the number of its generators. If $k(G) < 3$, then G is a group by Moufang's theorem. If $k(G) > 3$, then a subloop generated by any three elements must have order ≤ 27, so by the first part of the proof it must

H. P. Young

be a group. Therefore the associative law holds in G, so G is a group
and it is the unique abelian 3-group on four generators. The remaining
case, k(G) = 3, has been treated above. ∎

(17)   Corollary. The only c.M. 3-loop of order 27 is the abelian
3-group on three generators.

We will denote by "G_{81}" the unique c.M. 3-loop on three generators
that is not a group, and by "M_{81}" the corresponding Hall matroid.

Let M be any Hall matroid, and $G_e(M) = G$ the coördinatizing
loop of M based at e. The submatroid generated by a subset $A \subseteq G$
is the matroid $M \times cl_2(A)$. It is readily seen that the point-set
$A \cup \{e\}$ generates M if and only if A "generates" G in the loop
sense. Two lines (2-flats) L_1 and L_2 are parallel in M, written
$L_1||L_2$, if they are disjoint and contained in a common plane.

For any $h \in G$, $H = \{e,h,h^{-1}\}$ is a line of M, and the family
$\mathcal{L}_H$ of cosets xH, $x \in G$, is precisely the family of lines parallel
to H. The matroid structure implies that $\mathcal{L}_H$ partitions G, but in
general the members of $\mathcal{L}_H$ will not all be pairwise coplanar. Suppose
however, that h is in the center of G. Then for any distinct lines
x_1H, $x_2H \in \mathcal{L}_H$, $\{x_1,x_2,h\}$ generates a subgroup, and the corresponding
submatroid is an AG(3,3), whence $x_1H||x_2H$, because || is transitive
in an AG(3,3). Conversely, if $x_1H||x_2H$ for every pair of lines
x_1H, $x_2H \in \mathcal{L}_H$, then it can be shown that h is in the center of G.

A subset of lines of a Hall matroid $M = (E, \mathcal{H})$ that are pairwise
parallel and partition E will be called a central class of M.

H. P. Young

A central line is a member of a central class. It follows from the above
that a line L is a central line if and only if L is contained in the
center of the coördinatizing loop $G_x(M)$ for every $x \in L$.

In [3], Hall observes that M_{81} contains an AG(3,3) as a submatroid. In any Hall matroid, we shall call a 2-closed subset whose
lines and planes form an AG(3,3) a box. By (17), these are just the
2-closed subsets having cardinality 27. Using the present point of view
we may state more precisely the structure formed by the boxes in M_{81}.

(18)   Theorem. Let $\mathscr{L}'$ be the unique central class of the Hall
matroid M_{81}, and call any two distinct points x,y of M first
associates if $cl_2(\{x,y\}) \in \mathscr{L}'$, and second associates otherwise. Then
the boxes contained in M_{81} are the blocks of a partially balanced
incomplete block design on the two associate classes of points. The
number of blocks is 39, the number of blocks on every point is 13, and
$\lambda_1 = 13$, $\lambda_2 = 4$.

Proof. Let $\mathscr{L}'$ be the unique central class of M_{81}, and let $\mathscr{P}'$ be
the family of planes that contain an element of $\mathscr{L}'$. Then every plane
in $\mathscr{P}'$ contains exactly three members of $\mathscr{L}'$, and $(\mathscr{L}',\mathscr{P}')$ is an
affine triple system. For any point e, let H be the unique line in
$\mathscr{L}'$ containing e. Then H is the center of $G_e(M_{81})$ and it is a
straightforward matter to verify that $G' = G_e(M_{81})/H$ is the loop based
at H of $(\mathscr{L}',\mathscr{P}')$. Since G' is a group and $|G'| = 27$, it follows
that the 2-closed subsets of $(\mathscr{L}',\mathscr{P}')$ are the subspaces of an AG(3,3).
Let M' denote this AG(3,3), which is a rank 4 PMD.

H. P. Young

If S' is 2-closed in M', then $\Phi(S') = \bigcup_{L \in S'} L$ is 2-closed in M_{81}. If P' is any plane of M', then $\Phi(P')$ is 2-closed in M_{81} and has order 27, so it is a box. Let $\mathcal{B} = \{\Phi(P'): P' \text{ is a plane of } M'\}$. Now let K be any box in M_{81}. If K contains any member of $\mathcal{L}'$, then K is a disjoint union of elements of $\mathcal{L}'$, hence $K \in \mathcal{B}$. If K contains no member of $\mathcal{L}'$, let $e \in K$, and H the unique central line containing e. Then H is the center of $G_e(M_{81})$ and K is an order 27 <u>subgroup</u> of $G_e(M_{81})$ not containing H. Hence $G_e(M_{81}) = HK$ is a group, a contradiction. Hence $\mathcal{B}$ is precisely the set of boxes in M_{81}. Since $\mathcal{B}$ is in one-to-one correspondence with the planes of M', $|\mathcal{B}| = 39$. Further, every point of M_{81} (and also every pair of first associates) is contained in exactly one central line. Each central line of M_{81} is a point of M', and so is contained in exactly 13 planes of M'. Hence every point (and every pair of first associates) of M_{81} is contained in exactly 13 boxes. Finally, if x and y are second associates, let L_x and L_y be the distinct central lines containing x and y, respectively. Then every box of M_{81} containing $\{x,y\}$ contains $L_x \cup L_y$, hence there are four such boxes--the same as the number of planes of M' containing a given line of M'. ∎

(19)   <u>Corollary</u>. G_{81} contains exactly 13 subloops of order 27 (and they are all groups).

(19) is a special case of a theorem of Kulakoff (see [2], Ch. VI, Theorem 3.2), which states that: if G is a di-associative, non-cyclic, centrally nilpotent loop of odd prime power order p^n, then for $0 < m < n$ the number of subloops of G of order p^m is congruent to $1+p$ modulo p^2.

H. P. Young

(Every finitely generated commutative Moufang loop is centrally nilpotent-
see below.)

Given the existence of G_{81}, we may immediately construct a
c.M. 3-loop of order 3^n that is not a group, for any $n \geq 4$. Namely, we
take the direct product of G_{81} with n-4 copies of the cyclic group
of order 3. Nor are these the only possible constructions. Bruck ([2],
Ch. VIII) constructs an infinite c.M. 3-loop containing subloops on k
generators, for each $k \geq 4$, that are not factorable as G_{81} times a
group.

Next we show that every c.M. 3-loop has order a power of 3.

Where A, B, C are normal subloops of the commutative Moufang
loop G, (A,B,C) is defined to be the subloop generated by all
associators (a,b,c), where $a \in A$, $b \in B$, and $c \in C$. Then (A,B,C)
is normal in G. The <u>lower</u> <u>central</u> <u>series</u> of G is defined recursively
as follows.

$$G_0 = G$$

$$G_{i+1} = (G_i, G, G).$$

G is <u>centrally</u> <u>nilpotent</u> <u>of</u> (finite) <u>class</u> n if n is the least
integer for which $G_n = \{e\}$. The following fundamental theorem is
proved in [2].

(20)   <u>Theorem</u>. (Bruck-Slaby) If G is a commutative Moufang loop
generated by n elements then G is centrally nilpotent of class at
most n-1.

H. P. Young

Now let G be a finitely generated c.M. 3-loop with lower central series

$$\{e\} = G_m \subseteq G_{m-1} \subseteq \cdots \subseteq G_0 = G \; .$$

Then $G_{i+1} \triangleleft G_i$ for $0 \leq i \leq m-1$ and G_i/G_{i+1} is an elementary abelian 3-group, hence has order a power of 3. Since

$$|G| = \prod_{i=0}^{m-1} |G_i/G_{i+1}| \; ,$$

$|G|$ is a power of 3 and we have proved the following.

(21)   <u>Theorem</u>. Every Hall matroid has 3^n points for some integer n, and for every $n \geq 4$ there exists a Hall matroid on 3^n points.

4. Configuration Theorems

In this section we shall point out a fundamental configuration theorem that holds in any Hall matroid.

A set $\{w,x,y,z\}$ of four noncoplanar points in a Hall matroid, together with the six lines joining them, will be called a <u>tetrahedron</u>, and denoted by $T(w,x,y,z)$. $T(w,x,y,z)$ is said to be <u>singular</u> if $\{w,x,y,z\}$ generates an AG(3,3). The lines $cl_2(\{w \circ x, \; y \circ z\})$, $cl_2(\{w \circ y, \; x \circ z\})$, and $cl_2(\{w \circ z, \; x \circ y\})$ will be called the <u>braces</u> of T. The <u>midpoints</u> of the braces are the points $(w \circ x) \circ (y \circ z)$, $(w \circ y) \circ (x \circ z)$, and $(w \circ z) \circ (x \circ y)$.

(22)   <u>Theorem</u>. Let $T = T(w,x,y,z)$ be a tetrahedron in a Hall matroid M, and M' the submatroid generated by $\{w,x,y,z\}$.

H. P. Young

(i) If T is singular, then the midpoints of the braces
are identical.

(ii) If T is nonsingular, the three midpoints are distinct
and form a central line of M'.

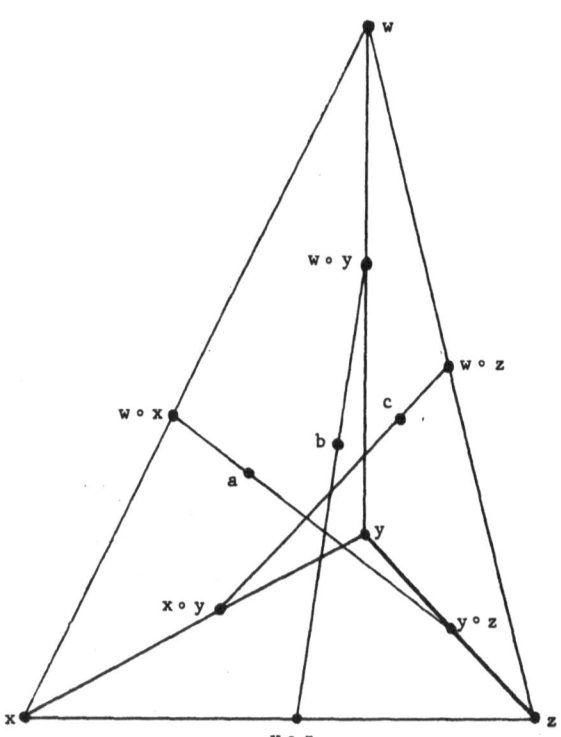

H. P. Young

Proof. The midpoints of the given tetrahedron may be represented in $G_w(M)$ by $a = x(yz)$, $b = y(xz)$, $c = z(xy)$. They will be collinear if and only if $a(bc) = 1$. Now $a(bc) = 1$ is equivalent to each of the following: $bc = a^{-1}$, $(bc)c = a^{-1}c$, and $bc^{-1} = a^{-1}c$. But

$$bc^{-1} = [(zx)y][z(xy)]^{-1} = (z,x,y)$$

and
$$a^{-1}c = [x(yz)]^{-1}[(xy)z] = (x,y,z)$$

Thus the collinearity of a,b,c follows from the skew-symmetry of the associator (9) in the loop $G_w(M)$.

Now let $L = \{a,b,c\}$. Then $w \notin L$, and $cl_2(\{w,a^{-1}c\}) = L'$ is the unique line containing w and parallel to L. But $a^{-1}c = (x,y,z)$ is in the center of the subloop of $G_w(M)$ generated by x,y,z. Hence L', and consequently L, are central lines of the submatroid generated by $\{w,x,y,z\}$. ∎

5. Conclusion

In this paper we have investigated the geometric properties of a class of perfect matroid designs, and shown how they relate to the algebra of commutative Moufang loops. This subject is developed more fully in [9]. It would be of great interest to find other algebraic structures that give rise to new classes of perfect matroid designs.

H. P. Young

References

1. Bruck, R.H.: "What is a Loop?" Studies in Modern Algebra,
   The Mathematical Association of America, 59-99 (1963).

2. _____ : A Survey of Binary Systems. Berlin-Heidelberg-
   New York: Springer (1966).

3. Hall, M.: "Automorphisms of Steiner Triple Systems".
   IBM Jour. Res. & Dev., 460-472 (1960).

4. _____ : "Group Theory and Block Designs". Proc. Internat.
   Conf. Theory of Groups, Austral. Nat. Univ. Canberra, 115-144 (1965).

5. Moufang, R.: "Zür Struktur von Alternativkörpern".
   Math. Ann. 110, 416-430 (1935).

6. Osborn, J. M.: "Loops with the Inverse Property".
   Pac. Jour. Math. 10, 295-304 (1960).

7. Young, P. and Edmonds, J.: "Matroid Designs".
   Jour. Res. Nat. Bur. Standards (to appear).

8. Young, H. P.: "Existence Theorems for Matroid Designs".
   Trans. AMS (to appear).

9. _____ : "Affine Triple Systems and Matroid Designs".
   Math. Z. (to appear).

EDITORIALE GRAFICA—ROMA